Frieder Johannes Koch

X-ray optics made by X-ray lithography

Process optimization and quality control

Schriften des Instituts für Mikrostrukturtechnik
am Karlsruher Institut für Technologie (KIT)
Band 36

Hrsg. Institut für Mikrostrukturtechnik

Eine Übersicht aller bisher in dieser Schriftenreihe
erschienenen Bände finden Sie am Ende des Buchs.

X-ray optics made by X-ray lithography

Process optimization and quality control

by
Frieder Johannes Koch

KIT Scientific Publishing

Dissertation, Karlsruher Institut für Technologie
KIT-Fakultät für Maschinenbau

Tag der mündlichen Prüfung: 6. April 2017
Referenten: Prof. Dr. Jan Gerrit Korvink, Prof. Dr. Franz Pfeiffer

Impressum

KIT Scientific Publishing

Karlsruher Institut für Technologie (KIT)
KIT Scientific Publishing
Straße am Forum 2
D-76131 Karlsruhe

KIT Scientific Publishing is a registered trademark
of Karlsruhe Institute of Technology.
Reprint using the book cover is not allowed.

www.ksp.kit.edu

Print on Demand 2017 – Gedruckt auf FSC-zertifiziertem Papier

ISSN 1869-5183
ISBN 978-3-7315-0679-9
DOI 10.5445/KSP/1000070029

X-ray optics made by X-ray lithography:
Process optimization and quality control

Zur Erlangung des akademischen Grades

Doktor der Ingenieurswissenschaften

der Fakultät für Maschinenbau
Karlsruher Institut für Technologie (KIT)

Genehmigte
Dissertation
von

Dipl. Phys. Frieder Johannes Koch

Tag der mündlichen Prüfung: 6. April 2017

Hauptreferent: Prof. Dr. Jan Gerrit Korvink

Korreferent: Prof. Dr. Franz Pfeiffer

"Un bon mot ne prouve rien."

Voltaire

Abstract

X-ray imaging is an indispensable tool in many areas from medicine over materials science to industrial quality control. Most applications rely on the differences in absorption between materials to generate image contrast. While this provides excellent results in many cases, the achievable contrast is severely limited for materials with similar absorption coefficients, such as different types of soft tissue. Grating based X-ray phase contrast imaging sets out to overcome this limit by also measuring the derivative of the X-rays phase and the scattering power of the sample. This can not only increase contrast for light materials but also provide information on the samples microstructure without the need to spatially resolve it. The implementation of this technique heavily relies on high quality gratings, which is a major challenge for fabrication, because the grating lines need to be both narrow and high, with a width in the single micrometer range and a height of up to a hundred times larger. These extreme aspect ratios can be achieved with X-ray lithography, where a photoresist is exposed with X-rays through a mask with gold absorber structures and the resulting polymer template is filled with metal.

This work aims to optimize the lithography process for higher quality X-ray optics, like gratings that can improve sensitivity in X-ray grating interferometry and compound refractive lenses with reduced aberrations. Two aspects of the lithographic process are in the focus: The thermomechanical properties of the photoresist template and the grating substrates. Key process parameters for improved grating quality could be identified by a combination of several techniques, most importantly tensile testing and gas pycnometry. The studies showed that the main factor for deformation of grating structures during the lithography process is not cure shrinkage of the resist but thermally induced stress, which can be mitigated by reducing the process temperatures. The introduction of new substrates allowed to better tailor gratings to low energy applications. These advances lead to a doubling of the sensitivity of a three grating interferometer at 27 keV and enabled the setup of an interferometer at only 8.3 keV design energy.

The optimization of lens fabrication first requires a precise metrology to quantify aberrations. As grating based phase contrast imaging provides an ideal tool for this purpose, a highly sensitive grating interferometer was designed and set up at the synchrotron ESRF. It allowed the quantitative comparison of X-ray lenses from different manufacturing techniques, highlighting the individual advantages and disadvantages, and specifically provided important feedback for lithographic lens fabrication. A gradient in the focal length of the lenses could be detected and traced back to the same thermally induced deformation that compromised grating quality. The process adaption that was shown to increase grating quality is therefore expected to also increase lens quality and in general the quality of any lithographic structure made from this photoresist.

Kurzfassung

Die Röntgenbildgebung ist als Werkzeug in vielen Bereichen wie Medizin, Materialwissenschaften oder der industriellen Qualitätskontrolle nicht mehr wegzudenken. Die meisten Anwendungen nutzen dabei die Unterschiede in der Absorption verschiedener Materialien um Bildkontrast zu generieren. Während dies für eine Reihe von Proben exzellente Ergebnisse liefert, ist der Kontrast zwischen Materialien mit ähnlichen Absorptionskoeffizienten stark limitiert, wie etwa im Fall verschiedener Weichgewebe-typen. Die gitterbasierte Röntgen-Phasenkontrastbildgebung versucht diese Limitierung zu überwinden indem neben der Absorption auch die differentielle Phase der Strahlung und die Streustärke der Probe gemessen wird. Das kann nicht nur den Kontrast zwischen Materialien aus leichten Elementen erhöhen, sondern auch Informationen über die Mikrostruktur einer Probe liefern ohne dass diese räumlich aufgelöst werden muss. Die Implementierung dieser Technik benötigt Gitter von hoher Qualität, die eine besondere Herausforderung für die Mikrofertigung darstellen, da die Gitterlinien gleichzeitig schmal und hoch sein müssen, mit einer Breite im einstelligen Mikrometerbereich und einer um einen Faktor bis zu 100 größeren Höhe. Diese extremen Aspektverhältnisse lassen sich mit der Röntgenlithographie erreichen, bei der ein Photoresist durch eine Maske mit Goldabsorbern hindurch mit Röntgenstrahlung belichtet wird und die resultierende Polymermatrix anschließend mit Metall befüllt wird.

Diese Arbeit zielt darauf, den Lithographieprozess zu optimieren um durch die Herstellung von Gittern mit höherer Qualität die Sensitivität von Röntgeninterferometern zu verbessern und die Aberrationen von Röntgenlinsen zu verringern. Zwei Aspekte des Prozesses sind dabei im Fokus: Die thermomechanischen Eigenschaften der Photoresistmatrix sowie die Substrate für die Herstellung. Die Schlüsselrolle einiger Prozessparameter konnte durch die Kombination verschiedener Analyseverfahren, insbesondere Zugversuche und Gaspyknometrie, identifiziert werden. Die Versuche zeigten, dass nicht der Volumenschwund bei der Vernetzung, sondern thermisch generierter Stress die Hauptursache für die Verformung von Gitterstrukturen während des Prozesses ist. Dieser kann durch Verringerung der Prozesstemperaturen reduziert werden. Die Einführung von neuen Gittersubstraten erlaubte weiterhin eine bessere Anpassung der Gitter an Anwendungen bei niedriger Photonenenergie. Die Fortschritte führten zu einer Verdoppelung der Sensitivität eines Drei-Gitter-Interferometers mit einer Designenergie von 27 keV und erlaubten den Aufbau eines Interferometers mit einer Designenergie von nur 8,3 keV.

Die Optimierung der Herstellung von Linsen erfordert zunächst eine hochpräzise Metrologie, um Aberrationen zu quantifizieren. Da die gitterbasierte Phasenkontrastbildgebung ein ideales Werkzeug für diesen Zweck darstellt, wurde ein Gitterinterferometer mit besonders hoher Sensitivität konzipiert und an der Synchrotronstrahlungsquelle der ESRF aufgebaut. Es erlaubte einen quantitativen Vergleich von Linsen aus verschiedenen Herstellungsverfahren und stellte deren individuelle Vor- und Nachteile heraus. Im Besonderen lieferte es wichtige Informationen für die lithographische Linsenherstellung, indem ein Gradient in der Brennweite über der Apertur detektiert wurde, dessen Ursprung sich auf dieselbe thermisch induzierte Verformung zurückführen ließ, die auch die Gitterqualität beeinträchtigte. Es wird daher erwartet, dass die Adaptierung des Prozesses, die bei der Gitterherstellung zu höherer Strukturtreue führte, auch zu kleineren Linsenfehlern führt und sich generell in einer höheren Strukturqualität von röntgenlithographischen Strukturen niederschlägt.

Contents

Abstract I

Kurzfassung III

List of abbreviations VII

Acknowledgement IX

1 Introduction 1

2 Theoretical background 7

2.1 X-rays 7
2.1.1 X-ray sources 7
2.1.2 Coherence 9
2.1.3 Interactions with matter 10
2.1.4 Optical elements 13
2.2 X-ray phase contrast imaging 16
2.2.1 The Talbot effect 17
2.2.2 Grating based phase contrast imaging 19
2.2.3 Sensitivity 25
2.2.4 Applications 26
2.2.5 Grating requirements 27

3 X-ray lithography 31

3.1 Substrates 32
3.2 Mask and exposure 33
3.3 Development and drying 36
3.4 Electroplating 37
3.5 Photoresist in X-ray lithography 38
3.5.1 Composition of mr-X 39
3.5.2 Layer preparation 40
3.5.3 Post exposure bake 41

4 **Photoresist Characterization** 43

4.1 Standard parameters 43
4.2 Thermal analysis by differential scanning calorimetry 45
4.3 Mechanical analysis by tensile testing 47
4.3.1 Variation of exposure parameters 50
4.3.2 Variation of storage time and temperature 55
4.4 Measurement of the volumetric shrinkage during crosslinking 59
4.4.1 Variation of storage time 64
4.4.2 Modified PEB for lithographic process 69
4.5 Summary of photoresist characterization 71

5 **Grating interferometry with low absorption substrates** 73

5.1 Polyimide 75
5.2 Graphite 77
5.3 Test samples and imaging results 78
5.3.1 Sensitivity 79
5.3.2 Low energy imaging 80

6 **Quantitative analysis of X-ray lenses** 83

6.1 Experimental setup 84
6.1.1 Focal length limits for quantitative analysis 87
6.2 Data analysis 89
6.2.1 Local lens aberrations 90
6.2.2 Global lens aberrations 92
6.3 Analysis of polymer compound refractive lenses 95
6.4 Finite element modeling of photoresist structures 102

7 **Conclusion and Outlook** 107

Appendix 111

A Phase retrieval 113
B Artefacts in angular deviation maps 115
C List of own publications 119

References 123

List of abbreviations

ANKA	Angströmquelle Karlsruhe
AR	Aspect Ratio
BPA	Bisphenol-A
CAD	Computer Aided Design
CRL	Compound Refractive Lens
CT	Computed Tomography
CTE	Coefficient of Thermal Expansion
DPCI	Differential Phase Contrast Imaging
(D)RIE	(Deep) Reactive Ion Etching
DSC	Differential Scanning Calorimetry
ESRF	European Synchrotron Radiation Facility
FEM	Finite Element Modeling
GBL	Gamma-Butyrolactone
IMT	Institute of Microstructure Technology
IMX	X-ray Imaging beamline at LNLS
KIT	Karlsruhe Institute of Technology
LIGA	Lithographie, Galvanik, Abformung
LNLS	Laboratorio Nacional de Luz Sincrotron
PAG	Photo Acid Generator
PEB	Post Exposure Bake

PEEK	Polyetheretherketone
PGMEA	Propylene glycol methyl ether acetate
PI	Polyimide
PMMA	Polymethylmethacrylate
TUM	Technische Universität München
UTS	Ultimate Tensile Strength

Acknowledgement

Like most bigger projects, be it in science or anywhere else, this one was not possible without the support of several others, some of which I want to mention specifically. First I would like to thank Dr. Jürgen Mohr for his guidance and support throughout my thesis. You were always available for questions and advice, took the time to read any text thoroughly and provide detailed feedback, which is not a small effort. Furthermore, I want to thank Prof. Jan Korvink and Prof. Franz Pfeiffer for taking over the duty of being the referees for this thesis. A large part of this work would not have been possible without the support of the Karlsruhe School of Optics and Photonics (KSOP) and the Karlsruhe House of Young Scientists (KHYS).

I want to thank the whole X-ray optics group of IMT for the wonderful working atmosphere and nice discussions, this includes current and former members: the "lens people" Arndt Last, Ottó Márkus, Elisa Kornemann, Sebastian Georgi and Felix Marschall as well as the "grating fans" Danays Kunka, Pascal Meyer, Vitor Vlnieska, Martin Börner, Max Amberger and Jan Meiser. Special thanks go to my dear office mates Tobias Schröter and Abrar Faisal for lots of things like keyboard gardening, assistance in trash basketball and admiring words for my clean desk. I also appreciate the great help I got from many other people at IMT, especially Alexandra Karbacher, Barbara Matthis, Christin Straus, Marie-Kristin Gamer, Julia Wolf, Martin Börner and Klaus Bade. I received a lot of help also from students, including but not limited to Martin Baumann, Chaitanya Lakkaraju, Adrian Schwarzenberger, Ibrahim Khalil and Andres Bello Hernandez.

Special thanks are due to Carsten Detlefs, who gave me the opportunity to work with him at the ESRF, I learned a lot from you! Throughout my work at the IMT, I had the opportunity to collaborate with many other researchers inside and outside of KIT, I specifically want to thank Lorenz Birnbacher, Marian Willner, Maite Ruiz-Yaniz, Julia Herzen and Franz Pfeiffer from TUM,

Alex Hipp from DESY/HZG, Fei Yang from EMPA, Sabine Engelhardt from KIT, Marie-Christine Zdora, Simone Sala and Irene Zanette from Diamond/UCL, Tunhe Zhou and Jenny Romell from KTH/Diamond, Jeff Wade and Claudio Ferrero from ESRF and Frank O'Dowd and Mateus Cardoso from LNLS.

Last but not least I want to thank my family and my wife for the constant support throughout not only my PhD work but also the long time preceding and hopefully following it.

Karlsruhe, January 2017

Frieder Koch

1 Introduction

The discovery of X-rays by Wilhelm Röntgen in 1885 [1] opened a whole new field of research that quickly sparked numerous applications that nowadays are indispensable in many fields like medical diagnostics, industrial quality control or security [2]–[5]. X-rays penetrate matter much better than visible light and can be used to get an image of the inside of samples without the need for mechanical slicing or opening. For the most part, today's imaging systems use the same experimental setup that Röntgen used in his laboratory: a source, based on accelerated electrons hitting a target, emits X-rays which then travel through the sample and are detected in a plane downstream. Image contrast is created by the absorption of photons in the sample and the resulting local intensity decrease on the detector. This yields excellent results when large differences in absorption coefficients are present in the sample, like bones in soft tissue. When little absorption is present, e.g. in very thin structures or material composed of light elements, the achieved contrast is limited. Materials with similar absorption coefficients are hardly distinguished in X-ray radiography; this is a major drawback especially in medical imaging, because the tissue that forms the inner organs, muscles and body fat is mainly composed of light elements and there is little difference in the absorption coefficients of different tissue types. This limitation can be overcome when using not the absorption, but the phase shift that X-rays undergo inside the sample. Both properties are described by the complex refractive index $n=1-\delta+i\beta$, where δ describes the phase shift and β the absorption. For light materials, δ can be three orders of magnitude larger than β, which means that a strong phase signal can be present even in cases of very faint absorption [6].

The phase cannot be directly measured by a detector, but various techniques have been proposed to access this information, the most popular being propagation based, coded aperture, and crystal or grating interferometry [7]. A particularly interesting method is grating based interferometry, which was first employed at synchrotron sources [8], [9], and transferred to conventional laboratory based sources in 2006 [10]. The technique relies on the Talbot effect, a self-imaging effect in periodic structures under coherent illumination [11]. A grating is used to create an interference pattern whose position is analyzed with the help of a second grating in case it cannot be directly resolved, e.g. when medical X-ray detectors with typical pixel sizes of tens to hundreds of micrometers are used. At X-ray tube sources, the necessary coherence is created by a third grating placed directly in front of the source. One of the gratings is then scanned in steps smaller than the grating period in the direction perpendicular to the grating lines and an image is recorded for every step. This procedure is done twice, once for reference and once with the sample in the beam path. The sample distorts the interference pattern, absorption reduces the mean value, refraction shifts the pattern to another stepping position and scattering reduces the amplitude of the pattern [12]. These three alterations of the interference pattern are calculated for each pixel and provide the absorption image, the differential phase contrast and the dark field or scattering image. The dark field image is of special interest, because it is sensitive to the microstructure of the samples on a scale that is not resolved by the imaging system, and can thus provide information altogether inaccessible to classical radiography.

The gratings are the key elements of this interferometric technique, they need to have periods in the range of a few micrometers, while the structure height should reach up to 200 µm and more for high photon energy. The resulting extreme aspect ratio presents a major challenge for microfabrication and can only be reached by highly specialized techniques. Aspect ratios of 100 and more in polymer and metal microstructures of the relevant sizes are possible with the X-ray LIGA technique (German acronym for *"Lithographie, Galvanik, Abformung"*, meaning "lithography, electroplating, molding") [13], [14]. In this technique, a photoresist is patterned by X-ray lithography using a mask fabricated by a mask-less technique such as electron beam lithography or direct laser writing. The photoresist used for grating fabrication is based on SU-8, a widespread resist in microlithography that undergoes a crosslinking reaction in the irradiated regions and becomes insoluble in most solvents [15]. After development, the resist forms the template for electroplating, during which the metal structure is grown in the resist matrix to form the grating structure.

The process can also be used to create other X-ray optical elements, in particular Compound Refractive Lenses (CRLs [16], [17]). CRLs have a lot of different applications, especially at synchrotron sources. Recent applications include illumination to create micro- to nanoscale beams, interferometry, and objectives for microscopy [18]–[20]. For all these, high manufacturing precision is of great importance, but difficult to achieve because of the extreme curvatures necessary because of the weak refractive power of the available materials, which in the case of X-ray lithography CRLs is the photoresist.

Both grating and lens fabrication require extreme precision of the lithographic structures and suffer from deformations of the photoresist that alter the structure geometry. The structure quality is highly dependent on the thermomechanical properties of the photoresist, which in turn are influenced by the process parameters like radiation dose, residual solvent content or crosslinking reaction temperature. In this work, the thermomechanical properties of the resist are studied under variation of the processing parameters, all with the goal of achieving optimal structure quality for both X-ray gratings and lenses for different applications. The presentation of these topics is divided into six chapters following this introduction. Chapter two provides the theoretical background of this work, starting with a short introduction to the creation of X-rays, their interaction with matter and X-ray optical elements. Following this, the principles of grating based X-ray phase contrast imaging are explained. The principle and state of the art of X-ray lithography as grating fabrication technique are described more detail in chapter three, and a special section is dedicated to the photoresist in X-ray lithography.

Chapter four describes the thermomechanical characterization of the photoresist. It contains a short overview of standard processing conditions and explains the main analysis techniques used for characterization: Differential scanning calorimetry (DSC), tensile testing and gas pycnometry. The results presented highlight the influence of radiation dose, waiting times between process steps and processing temperatures on the properties of the resist like Young's modulus, tensile strength, volumetric shrinkage during crosslinking and crosslinking reaction speed. A main result is that by reducing the temperature during the crosslinking reaction, one can reduce waviness in long grating lines and thus increase the structure quality of gratings. The stress generated in the structures by thermal expansion coefficient mismatches therefore appears greater than the stress induced by the chemical shrinkage.

Microscopic inspection of the structure quality of gratings is not sufficient to rate their performance in phase contrast imaging, because defects may be sub-surface and invisible to electron and light microscopy. The only way to fully assert the quality of a grating is by using it in an interferometric setup and determining the achievable fringe visibility and homogeneity. Chapter five shows experiments at an X-ray tube source, comparing gratings with equal period and structure heights but different substrates. It is shown that the substrate material does not influence structure quality significantly at moderate aspect ratio; switching from the standard silicon substrates to graphite and polyimide has no negative influence on the visibility, but the increase in transmission led to a decrease of the smallest resolvable refraction angle of a setup with given exposure time, thus increasing its sensitivity. Additionally, the gratings on low absorbing substrates enable grating interferometry experiments at low design energy, demonstrated by results from a synchrotron-based interferometer operating at 8.3 keV design energy.

The structure quality of X-ray lenses is not as easy to judge from the experiments they are intended for as in the case of gratings. The size and shape of the focus of a lens only provide limited information about structural defects. To analyze single refractive X-ray lenses, an interferometer was specifically designed and built up at the beamline ID06 of the European Synchrotron Radiation Facility (ESRF), which is presented in detail in chapter six. The interferometer was used for the analysis of refractive lenses made by Beryl lium imprinting and X-ray lithography and showed their respective advantages and disadvantages, namely a high shape accuracy but some localized defects in the Berylium lenses and high material homogeneity but a shape deviation that caused the focal length of the lens to vary over its aperture for the X-ray lithography lenses. This last effect is shown to stem from the deformation of the photoresist during the crosslinking reaction, and just as in the case of gratings, the dominant cause for deformation appears to be thermal stress during the processing instead of the stress introduced during the shrinkage in the crosslinking reaction. Chapter seven concludes this work with a summary of the results and an outlook for further experiments.

2 Theoretical background

2.1 X-rays

The following paragraphs describe the sources of X-rays in use today, the interaction of X-rays with matter and the optical elements used to manipulate X-ray beams.

2.1.1 X-ray sources

X-ray tubes

In most applications today, X-rays are produced using a principle very similar to what Röntgen used for his experiments. Electrons emitted from a cathode in an evacuated tube are accelerated in an electric field with a potential difference between a few and hundreds of kilovolt, and directed to a target, usually made out of a metal such as Copper, Molybdenum or Tungsten [21]. shows a schematic depiction of the principle.

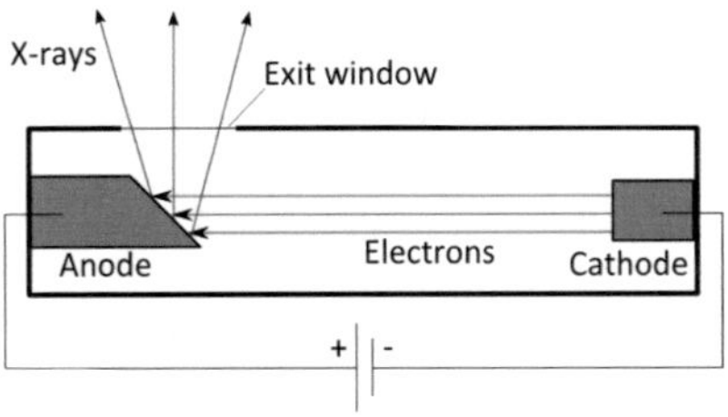

Figure 2.1: Principle of the X-ray tube

The accelerated electrons lose their kinetic energy mainly in two ways: Hitting electrons in the target material and transferring momentum to them, and decelerating in the field of an atomic nucleus. Both processes lead to the emission of X-rays. When an accelerated electron liberates an electron from the inner shell of a target atom, this void is filled with electrons from outer shells, and the energy difference between the two states is either radiated as an X-ray photon or transferred to another electron which is then liberated from the atom (Auger electrons). The difference of potential energies between the states of an atom, and therefore the energy of the emitted photon, is characteristic of the chemical element of the target. The second process, the deceleration in the electric field of the nucleus, is mostly governed by the initial energy of the electron. An accelerated or decelerated electric charge emits electromagnetic radiation called Bremsstrahlung. The photon energy corresponds to the amount of energy lost; the highest possible photon energy is therefore identical to the initial kinetic energy of the electron. The spectrum created by this process is continuous; the X-ray emission from these so-called tube sources is therefore polychromatic and highly dependent on the target material and acceleration voltage [21].

The output of a tube source is a cone beam. Apart from the radiation produced in the deceleration of electrons in the target material, a lot of heat is created, which limits the output intensity of a tube source. Several approaches exist to overcome this problem, a very important step forward was the introduction of the rotating anode principle, in which a constant rotational movement of the anode distributes the heat load over a larger area in the target and facilitates cooling [22]. These rotating anode tubes allow much higher output power; modern devices achieve around 100 kW with a source size of about 1 mm^2, see e.g. reference [23].

High resolution X-ray imaging requires small source sizes, and electron focusing optics are used in so-called micro focus tubes to narrow the impact zone on the target to a spot of only a few micrometers in diameter. With decreasing focal spot size, the local heat load increases, which in practice means that conventional micro focus tubes have a limited output power. A more recent approach replaces the solid target material with a liquid metal jet, which allows a substantial increase of the brilliance [24].

Synchrotrons and storage rings

Synchrotrons are large particle accelerators that use a series of bending and focusing magnets to put a charged particle on a closed trajectory. Radio frequency electric fields are used to accelerate the particles, and upon gaining energy, the fields in the bending magnets have to be ramped up synchronously (hence the name) in order to keep the particle trajectory constant. The change of direction in a magnetic field is an acceleration of the charged particle and thus gives rise to the emission of electromagnetic radiation. With electrons at relativistic speed, this effect provides the possibility to create intense X-ray beams that are called synchrotron radiation. The ANKA (Ångströmquelle Karlsruhe) synchrotron in Karlsruhe is an example for an accelerator specifically built to create synchrotron radiation; other examples are the European Synchrotron Radiation Facility (ESRF) in France, Spring-8 in Japan or the Advanced Photon Source in the USA. The radiation used for experiments at these facilities is referred to as synchrotron radiation for historical reasons, although it is usually not produced in a synchrotron, but in a storage ring. Storage rings have a similar layout as synchrotrons, but operate at constant energy and thus require acceleration of the particle to the final energy prior to injection, which is usually done with a series of linear acceleration and a booster synchrotron [25].

Synchrotron radiation can be produced in different devices, the simplest being the *bending magnets* used to keep the particle on its closed trajectory. The Lorentz force exerted on the electrons in the magnetic field accelerates the charges which consequently radiate photons. Because the particles are moving at relativistic speed, the photons are emitted in a narrow cone pointed in the momentary direction of the particle, tangential to the orbit. The need for higher intensity, brilliance and monochromatic illumination has led to the introduction of more sophisticated devices for the production of synchtrotron radiation, namely wigglers and undulators, which are based on alternating magnetic fields [26].

2.1.2 Coherence

For X-ray imaging, especially for high resolution or phase contrast imaging, the spatial coherence of the beam is a very important property. Essentially, spatial coherence is a measure of how well the wavefront resembles a wavefront created by a theoretical point source. The transversal coherence length

ξ of a monochromatic source can be calculated from the wavelength λ, the distance l of the source to the observer and the source size s by [26]

$$\xi = \frac{\lambda \cdot l}{s}. \tag{2.1}$$

By definition, the transversal coherence length of a plane wave and of a wave from a point source is infinite. Apart from astronomical objects, the most coherent real X-ray sources available are synchrotron sources and X-ray free electron lasers, whose transversal coherence length can reach hundreds of micrometers [27], while the typical transversal coherence of X-ray tube sources is only several hundred nanometers.

2.1.3 Interactions with matter

In the following section, a brief introduction to the basic concepts of the interaction of X-rays with matter will be given. A more thorough approach can be found in Refs. [26], [28]. Figure 2.2 shows a schematic depiction of what happens on a macroscopic level when an X-ray passes through an object. Wave B propagates undisturbed, while wave A passes an object. In the object, the propagation speed differs from the surrounding, which means that the wavelength is changed, and after the object there is a difference $\Delta\varphi$ between the phase of A and B. Moreover, the intensity of A is attenuated in the object, leading to smaller amplitude. Both effects can be described with the complex refractive index n:

$$n(x,y,z) = 1 - \delta(x,y,z) + i\beta(x,y,z) \tag{2.2}$$

The quantities δ and β in equation (2.2) are real numbers, and for X-rays, both are also positive; δ is called the refractive index decrement, and β relates to the linear attenuation coefficient μ (see following section). If we describe the X-ray in terms of the scalar wave field E(x,y,z), its propagation along the direction z through an object of thickness Z can be written as

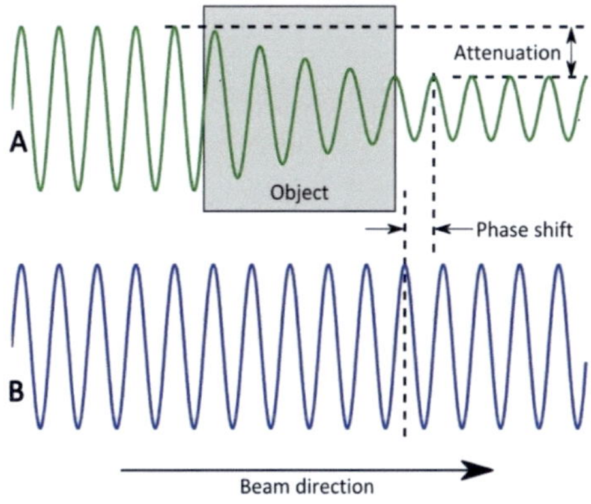

Figure 2.2: Schematic representation of attenuation and phase shift introduced to a beam by an object

$$\begin{aligned} E(x,y,Z) &= E(x,y,0)\cdot exp\left(ik\int_0^Z n(x,y,z)dz\right) \\ &= E(x,y,0)\cdot exp\left(ik\int_0^Z (1-\delta(x,y,z))dz\right)\cdot exp\left(-k\int_0^Z \beta(x,y,z)dz\right). \end{aligned} \tag{2.3}$$

Here, $k = 2\pi/\lambda$ is the wavenumber and λ is the wavelength of the radiation. The second line in equation (2.3) shows that the effect of the object on the wave field can be separated into a term only affecting the phase (phase shift) and one only affecting the intensity (attenuation). Both are described in more detail in the following.

Attenuation

The intensity I of a wave is defined as the square of its amplitude $I(xyz) = |E(x,y,z)|^2$. Applying equation (2.3), the transmission of an object is given by:

$$T(x,y,z) = \frac{I(x,y,z)}{I(x,y,0)} = \frac{|E(x,y,z)|^2}{|E(x,y,0)|^2} = exp\left(-2k\int_0^z \beta(x,y,z)dz\right).$$

This is equivalent to the Beer-Lambert law $T(x,y,z)=exp\left(-\int\mu(x,y,z)dz\right)$ with the *linear attenuation coefficient* $\mu=2k\beta$. In the energy region relevant to this work, i.e. the energies used for X-ray lithography and imaging, three physical mechanisms are responsible for attenuation of the beam:

- Photoelectric absorption
- Elastic scattering (Rayleigh scattering)
- Inelastic scattering (Compton scattering)

In light materials and for low photon energies, the dominant process is photoelectric absorption. The photon is completely absorbed and its kinetic energy is transmitted to an electron. If the transmitted energy exceeds the binding energy of the electron to the atomic nucleus, it is liberated from the atom, leaving behind a vacancy in the orbital state it occupied. This vacancy will be filled with an electron from a state with a higher energy, and the excess energy can be emitted as a photon of the characteristic energy described in section 2.1.1. The other possibility is the transfer of this energy to another electron, which is also emitted from the atom. This process, called the Auger effect, is dominant for light elements.

The cross section of the photoelectric absorption has a strong dependence on the photon energy E and the atomic number Z of the absorber, the relationship can be approximated as

$$\sigma_{pe} \propto \frac{Z^5}{E^{3.5}}. \tag{2.4}$$

If the photon energy is very close to binding energy of an electronic state of the absorber, the absorption cross section becomes much larger and relation (2.4) is not a good approximation any more. This step in σ_{pe} is called an absorption edge, and corresponding to the orbital angular momentum of the electronic state involved, the edges of an element are referred to as K-, L-, M-edge etc.

Elastic scattering, i.e. scattering without transfer of energy, is a process mainly occurring at low photon energy, it can be explained using only the classical electromagnetic theory. The scattered photon has a defined phase shift of π with respect to the incoming photon, in other words, the process is coherent. At X-ray energies higher than 10-20 keV, inelastic scattering becomes more important, and at high energies, it is the dominant interaction process of X-rays with matter. Opposed to elastic scattering, there is no relation between the phase of the scattered and the initial photon, the process is incoherent. In literature, the terms Compton, inelastic and incoherent scattering are used to refer to the same process.

The total attenuation cross section is calculated as the sum of the contributions from the three processes: $\sigma_{tot} = \sigma_{pe} + \sigma_{coh} + \sigma_{incoh}$.

Phase shift

The phase $\Phi_{obj}(x,y,z)$ of the wave E(x,y,z) after passing through an object of thickness Z is given by the imaginary part of the exponent in equation (2.3) added to the initial phase. In many cases, it is interesting to know the phase shift relative to an undisturbed reference ray travelling the same geometrical distance. A simple subtraction yields

$$\Delta\Phi = \Phi_{ref} - \Phi_{obj} = k\int_0^Z dz - k\int_0^Z \left(1-\delta(x,y,z)\right)dz = k\int_0^Z \delta(x,y,z)dz.$$

For a homogeneous object of thickness Z, the phase shift is thus $\Delta\Phi = k\delta Z$.

It is interesting to note that δ is positive for X-rays, the real part of the refractive index is thus smaller than 1. At first sight, this seems like a violation of the principle that no information can travel faster than the speed of light *c*, but one has to consider that the velocity that determines the refractive index is the phase velocity, but information can only travel with the group velocity, which is smaller than *c* [29].

2.1.4 Optical elements

Over the course of the decades, many different devices for the manipulation of X-ray beams have been developed. These optical elements can be grouped by the physical process they are based on:

- Absorption
- Diffraction
- Reflection
- Refraction

Absorbing optics include filters to reduce intensity or to shape the spectrum of a polychromatic source, slits and apertures are used for a variety of purposes. Another type of absorbing optics are the shadow masks used in X-ray lithography, see section 1, as well as the source and analyzer grating in a Talbot-Lau interferometer, see section 2.2.2.

Two types of diffracting optics are commonly used with X-rays: Fresnel zone plates and diffraction gratings. Fresnel zone plates, named after Augustin-Jean Fresnel who first described the principle [30], are used for focusing X-rays; they consist of concentric rings with a thickness that shrinks from the center to the outer part. The width of the outermost ring determines the diffraction limit when using a zone plate [31], so the image resolution that can be obtained is limited by the fabrication technique used for the zone plate. State of the art fabrication techniques yield feature sizes down to 12 nm, but due to limited zone height, these high resolution zone plates reach very low efficiency and are not well suited for high energy applications [32]. The other type of diffractive optics, the grating, will be described in more detail in section 2.2.2.

As a consequence of the refractive index being smaller than one, the phenomenon of total external reflection occurs at air-material interfaces. It is the same principle as total internal reflection used e.g. in fiber optical elements. However, as the difference between the refractive index of air and any material is typically in the order of 10^{-6}, the critical angle for total external reflection $\theta_c = arcsin(n_1 / n_2)$ is very close to 90°. For convenience, it is thus usually given not measured from the normal to the interface as it is customary in visible light optics, but from the tangent to the interface. Angles in total external reflection geometries then are very small and the geometry is typically referred to as "grazing incidence". An example of a type of X-ray optic using this principle is the Kirkpatrick-Baez-mirror [33], that uses two curved mirrors to create a focal spot downstream. A distinct advantage of mirror optics over other types is their inherent achromaticity, disadvantages are the complicated alignment, and the extreme manufacturing precision needed on a large area which makes the mirrors very expensive.

The fourth type of optics is refracting, i.e. lenses that are based on the refraction of a ray at the interface between two materials. The focal length f of a lens depends on the parabola apex radius R and the refractive index decrement δ, as well as the number N of elements it consists of. As δ is very small, a single refractive lens, even with extreme curvature, has a weak refractive power; many refractive elements have to be combined to achieve a smaller focal length. Such a lens is called a compound refractive lens (CRL [16]). If all elements have the same radius of curvature, the effective focal length is given by

$$f = \frac{R}{2\delta N}. \tag{2.5}$$

As the refractive index is smaller than 1, a focusing lens for X-rays is biconcave, instead of biconvex as it is the case for visible light. The principle of a CRL is depicted in Figure 2.3.

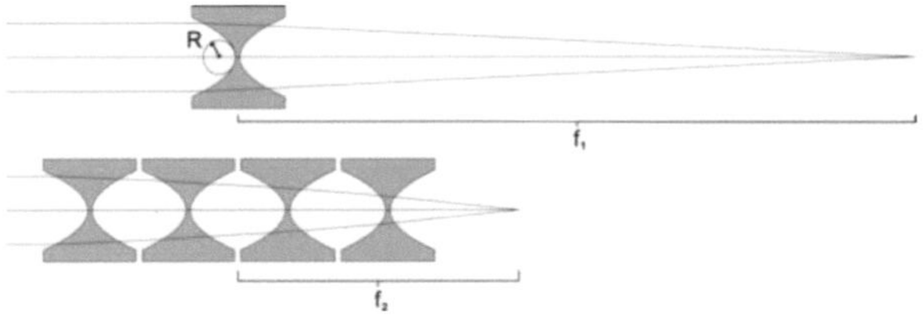

Figure 2.3: Principle of a compound refractive lens for X-rays with parabola apex radius R and focal lengths f_1 and f_2, from [34]

With δ of the order 10^{-6}, it is obvious from equation (2.5), that for a focal length below a meter, R has to be in the micrometer range, requiring microfabrication techniques with high precision. In addition, the material of the lens should be highly transparent for X-rays, and consequently light materials are used for lens fabrication, such as beryllium, aluminum, silicon or polymers. In ductile materials like Aluminum and Beryllium, lenses can be fabricated individually by imprinting [35]. This provides the possibility of directly fabricating point focus lenses, but is limited in terms of minimal apex radius. Current commercially available lenses have a minimum apex radius of 50 µm [36]. To get to smaller focal length, other manufacturing techniques have to be employed. Lenses can be fabricated in silicon by reactive ion etching (RIE) [37], allowing much smaller apex radius, or in polymer by X-ray lithography, a process which will be described in more detail in section 1. Lenses with small apex radius are limited in aperture, because with distance from the optical axis, the thickness of the lens material increases and the transmission decreases.

All refractive lenses for X-rays suffer from chromatic aberration, which is very hard to correct for because of the small differences in refractive index between different materials. They are thus mainly suited for monochromatic illumination.

2.2 X-ray phase contrast imaging

Although the images obtained in hospitals or non-destructive laboratories nowadays are of much higher quality than what Röntgen was able to achieve, they essentially use the same principle, only exploiting the absorption, and thus the particle nature of X-rays. As pointed out in section 2.1.3, the absorption of a photon mainly depends on its energy and the atomic number of the absorber. For a given spectrum, the contrast that can be achieved between materials of similar atomic number, such as different types of soft tissue, is therefore limited.

Over the years, several methods have been developed to overcome this limit by also measuring the phase of X-rays, which is not directly accessible by detectors, as they only measure the intensity of a beam. The contrast in the phase shift can be as much as a factor of 1000 higher than in absorption for light materials [6]. The first successful attempt was an interferometer based on a single silicon crystal [38], other approaches used a crystal as analyzer [39]. Today, a widespread method is based on the propagation of the wave, also called in-line phase contrast [40], [41]. All crystal based methods require highly spatially coherent and or monochromatic illumination and can only be applied efficiently at synchrotron sources. The requirements of propagation based phase contrast are less strict; it is tolerant to divergent polychromatic beams with moderate coherence and can be realized at microfocus X-ray tubes, for example. Another interesting technique uses the displacement of a speckle pattern generated by a random phase modulator, such as sand paper, to detect the phase shift of the sample. This technique, too is tolerant to polychromatic radiation and requires only the grade of coherence that is available at microfocus laboratory sources [42]. To be able to access the higher flux that can be provided by a rotating anode source with less coherence, one has to switch to a setup called a Talbot-Lau interferometer [10], whose basics will be explained in the following section.

2.2.1 The Talbot effect

In 1836, Henry Fox Talbot published his findings that a periodic structure, when illuminated through a slit, produced an image of itself in discrete distances downstream [11]. This phenomenon is now known as the Talbot effect, and analytical calculations show that when illuminated by a plane wave, any periodic structure will produce such a self-image at distances z_n determined by

$$z_n = n \cdot \frac{2a^2}{\lambda}. \tag{2.6}$$

In this equation, $n = 1,2,3,...$ is referred to as the Talbot order, a denotes the period of the structure, and lambda refers to the wavelength of the illumination. The interference pattern between these distances can be complicated and depends strongly on the properties of the periodic structure. It is often called the Talbot carpet. A particularly interesting case of the periodic structure is a pure phase object, called a phase grating. In the idealized case, such an object absorbs or scatters no photons, but only introduces a box-like periodic phase shift to the wavefront. Depending on the amount of this phase shift, there are distances between the Talbot orders, at which this phase contrast is entirely transformed into intensity and can thus be measured by a detector.

Figure 2.4 shows such a Talbot carpet for a phase grating with a duty cycle (the ratio of the width of the phase shifting lines to the grating period) of 0.5 and a phase shift of $\pi/2$ up to the second Talbot order. As the structure is a pure phase object, there is no intensity contrast at the Talbot distances, but at $0.25 \cdot z_n$ and then recurring every $0.5 \cdot z_n$, all the intensity is concentrated in one half of the grating period.

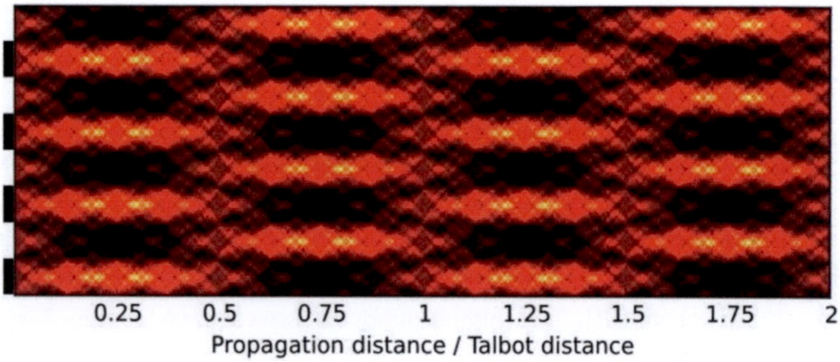

Figure 2.4: Simulated Talbot carpet for a $\pi/2$ shifting phase grating under plane wave illumination

If the phase grating is changed to induce a phase shift of π instead of $\pi/2$, the Talbot carpet changes its shape quite strongly, as shown in Figure 2.5. There are still fractional Talbot orders at which the phase contrast is converted to pure intensity contrast, but instead of two, there are now 8 such distances before the first full Talbot distance. Another difference is that at these fractional distances, the period is half the period of the phase grating.

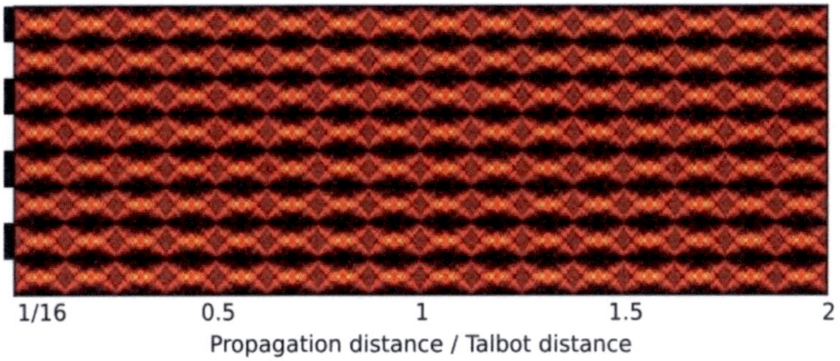

Figure 2.5: Simulated Talbot carpet for a π-shifting phase grating under plane wave illumination

If one introduces an object into the beam, the Talbot carpet is disturbed, the simple case of a wedge shaped pure phase object is depicted in Figure 2.6 for a π/2 phase grating. The refraction by the object shifts the lines of the Talbot carpet to different positions, and the analysis of these positions can yield the phase shift introduced by the object.

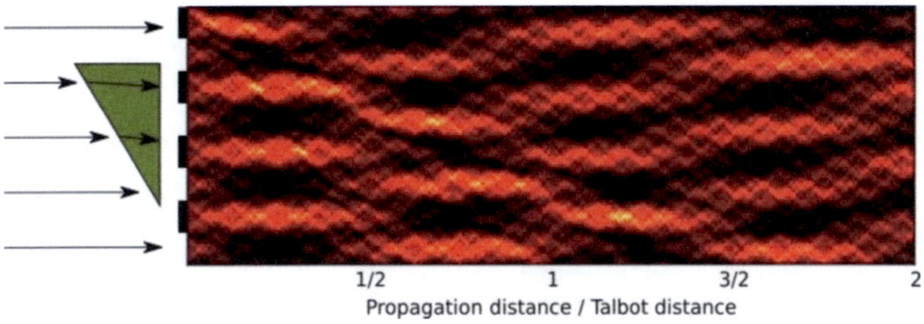

Figure 2.6: Simulated Talbot carpet with a wedge shaped pure phase object in the beam path immediately before the grating

2.2.2 Grating based phase contrast imaging

To estimate the dimensions of a Talbot interferometer and the necessary gratings, we look back at equation (2.6) on page 17. The wavelength of X-rays is roughly in the order of 1 Å, which is $1 \cdot 10^{-10}$m. A reasonable demand is a setup length that fits on standard optical tables, so the Talbot distance should be in the meter range, which leaves the grating period in the order of $10^{-5}-10^{-6}$ m, meaning a few micrometers. To resolve a grating period like this, high resolution cameras are needed, which necessarily come with a small field of view. Detectors in medical imaging or other applications that need a larger field of view come with pixel sizes above 50 µm, making it impossible to resolve the interference pattern. The solution for this problem is the introduction of a second grating, whose period matches exactly the one of the interference pattern. This grating needs to have absorbing lines and acts as a shadow mask at a fractional Talbot distance, it is commonly referred to as either analyzer grating or G2, with the phase grating being G1. The detector is placed directly behind the analyzer grating, the setup is shown in Figure 2.7.

Figure 2.7: Setup of a Talbot interferometer with phase and analyzer grating

If the lines of the analyzer grating are exactly at the positions of the intensity maxima, no photons reach the detector in the ideal case and displacing the analyzer grating by half a period will transmit the full intensity. To access the whole interference pattern, one of the gratings is moved in steps smaller than a period in the direction x_g perpendicular to the grating lines and the optical axis. The intensity is recorded for each position, yielding a so-called stepping curve for each detector pixel. For an ideal phase and absorption grating with the perfect relative distance and plane wave illumination, this curve would have a triangular shape, as it is the convolution of two rectangular intensity profiles. In practice, due to the finite source size and grating imperfections, this curve is smeared out and can usually be modeled by a sinus curve. Figure 2.8 shows two model stepping curves, one reference curve and one with a sample.

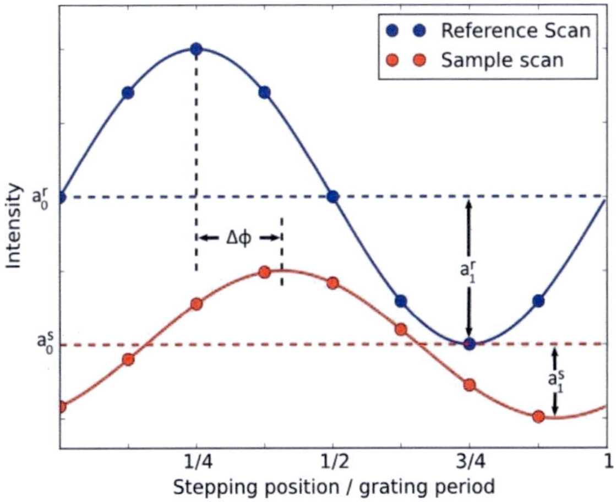

Figure 2.8: Schematic of a reference stepping curve and a stepping curve with sample in beam

The period of the stepping curve is given by the period of the interference pattern, so three parameters are left to describe the curve: The mean value a_0, the phase φ and the amplitude a_1. All three parameters are influenced when a sample is put in the beam. The absorption in the sample will reduce the mean intensity, refraction in the sample will displace the interference pattern and therefore change the phase, and scattering will reduce the amplitude of the stepping curve. While the contributions to the amplitude can be complicated, it is important to note that the mean value and the phase are independent of each other, in particular the phase contains no contribution from absorption [9]. An important parameter for the analysis of the stepping curve is its contrast, which is usually quantified by the visibility V:

$$V = \frac{I_{max} - I_{min}}{I_{max} + I_{min}}. \tag{2.7}$$

The higher the visibility, the more robust the extraction of the curve parameters, which in turn results in a higher contrast to noise ratio in the final images. A stepping curve can be analyzed by doing a sinusoidal fit for each pixel, but a much faster approach in terms of calculation time uses a discrete Fourier transform [43]. The stepping curve intensity $I(x_g)$ can be written as a Fourier series:

$$I(x_g) = a_0 + \sum_{k=1}^{\infty} a_k \cdot cos(k \cdot x_g + \phi_k) \approx a_0 + a_1 \cdot \cos(\mathrm{k} \cdot x_g + \phi_1) \tag{2.8}$$

It is easily seen that with this notation, the visibility becomes

$$V = \frac{a_1}{a_0}.$$

Note that most implementations of the discrete Fourier transform in modern programming languages such as Python or Matlab return the coefficients in a format where the a_1 does not correspond to the amplitude, but twice the amplitude of the fundamental frequency. The visibility then becomes $V = 2 \cdot a_1 / a_0$. When the reference stepping curve (superscript r) and the curve with sample (superscript s) are analyzed, three image modalities can be extracted. First, the transmission T, as measured in a conventional radiography, is given by

$$T = \frac{a_0^s}{a_0^r}. \tag{2.9}$$

The shift $\Delta\phi$ of the interference pattern is calculated as $\Delta\phi = \phi_{1r} - \phi_{1s}$, and the differential phase $\partial\Phi / \partial x$ of the wavefront can be calculated from it using the period p_2 of the interference pattern, the inter-grating distance d and the wavelength λ of the illumination:

$$\frac{\partial \Phi}{\partial x} = \Delta \phi \frac{p_2}{\lambda d}. \tag{2.10}$$

The refraction angle α is related to the differential phase as

$$\alpha = \frac{\lambda}{2\pi} \frac{\partial \Phi}{\partial x}. \tag{2.11}$$

The third imaging modality is the loss of visibility, often called the dark field D because of its relation to small angle scattering [12]

$$D = \frac{V_s}{V_r}.$$

Within this work, this imaging modality will always be referred to as dark field, as there is no overlap with other techniques that use the term with a different meaning.

Figure 2.9 shows an example to illustrate the differences between the imaging modalities. The sample is a plastic keychain figure with a textile cape, part B of the figure shows the X-ray transmission image, which clearly shows the metal screw inside the figure used to attach it to a keychain. The contrast between this screw and the plastic is excellent because of the large differences in absorption coefficients. The differential phase image in part C shows nicely the structure details and gives an idea of the three-dimensional shape. The textile cape of the figure is only visible in the dark field image, as it is very thin and produces only very little absorption. However, its fibrous fabric scatters the photons and it therefore reduces the visibility, making it appear very clearly in this modality.

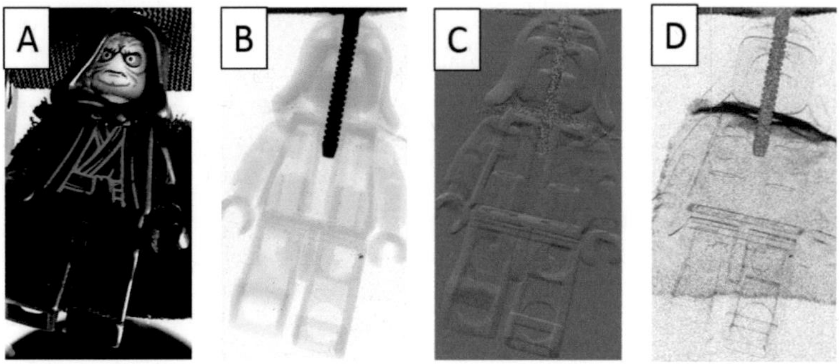

Figure 2.9: Example for the modalities of grating based X-ray phase contrast imaging. A: Photograph of the sample, a plastic keychain figure, B: X-ray transmission image, C: Differential phase, D: Dark field image. Taken from [44].

In order to observe the Talbot carpet up to the first Talbot order, the spatial coherence ξ (see section 2.1.2) has to be greater that the period p_1 of the phase grating. A standard X-ray tube does not meet this requirement, and so another absorption grating, called G0, must be introduced between source and phase grating. This grating produces an array of slit sources, which individually provide enough coherence to observe the Talbot effect, but are mutually incoherent; the loss of spatial resolution of the sample through blurring is still determined by the original source size. In order for the individual slits to contribute constructively to the interference pattern in the detector plane, their period p_0 has to satisfy the geometrical condition [10]

$$p_0 = p_2 \cdot \frac{l}{d}. \tag{2.12}$$

Again, p_2 refers to the period of the interference pattern and d is the inter-grating distance. The quantity l now denotes the distance from G0 to the phase grating. The transverse coherence now becomes $\xi = \lambda \cdot l / (1 - \gamma_0) \cdot p_0$, with γ_0 as the duty cycle of the source grating.

An interferometer that uses a source grating to provide the necessary coherence is called a Talbot-Lau-Interferometer, or often a three-grating interferometer, although still only one of the three gratings actually acts as a diffraction grating. It was first realized in 2006 [10], and allows the use of virtually any kind of X-ray source for the formation of X-ray phase contrast images, which makes it a very promising technique for industrial applications that rely on accessibility. Aside from the use with X-rays, such an interferometer can also be used with common neutron sources [45]. Interferometers can be realized in a wide variety of geometries, depending on the priorities, e.g. compactness, high sensitivity or high spatial resolution. Any choice of a grating parameter affects all the other parameters, a full set of the formulas required for the setup of a new interferometer can be found in reference [46], a web-based tool for the calculation is freely available at [47].

2.2.3 Sensitivity

The key parameter for evaluation of the performance of a given interferometer setup is the sensitivity. It is usually expressed by the smallest detectable refraction angle α_{min}. The smaller α_{min}, the higher the sensitivity. From equations (2.10) and (2.11), the refraction angle can be expressed in terms of interferometer parameters

$$\alpha = \frac{p_2}{2\pi d}\Delta\phi.$$

It is directly obvious that for a high sensitivity, the grating period should be chosen as small as possible and the inter-grating distance should be maximized. The grating period is limited by fabrication issues, and the inter-grating distance is practically limited by the available flux and the coherence of the source. Apart from these geometrical factors, the angular sensitivity is determined by the noise σ_ϕ of the measured signal $\Delta\phi$ [48], which for a photon counting detector can be expressed in terms of the visibility V and the number of photons N:

$$\sigma_f = \frac{\sqrt{2}}{V\sqrt{N}} \tag{2.13}$$

In general, the visibility decreases with increasing inter-grating distance, so a compromise must be made to maximize the sensitivity.

2.2.4 Applications

Both the differential phase contrast and the dark field contrast provide complimentary information on the sample. As they are of different nature, they also have different applications. The phase signal can for example be used for quantitative analysis of light materials that hardly show any contrast in absorption [49] and offers the possibility to detect the cartilage within a joint in an human hand [50]. It is also a useful tool for the characterization of X-ray optical elements, such as mirrors and lenses [51], [52], and its application as such a tool is a main part of the experimental section of this work, see chapter 0.

The dark field signal is sensitive to scattering; it emphasizes interfaces between materials and depends strongly on the microstructure of the sample. This is particularly interesting, because it means that the signal is sensitive to details of the sample that are well below the systems spatial resolution. An example of a strongly scattering structure in mammals is the lung, with its millions of micrometer-sized air chambers called alveoli. Some lung diseases like emphysema change the microstructure of the lung, effectively reducing its surface area and negatively affect its function. While such changes are almost invisible in a conventional transmission radiograph, they have a much stronger effect on the dark field signal and could thus potentially be detected at a much earlier stage [53]. Other interesting application are the detection of microcalcifications in mammography, as these are indicators for breast cancer [54], [55] and the classification of renal stones from radiographs [56].

In materials science, the dark field is especially useful for the investigation of porous materials. One example is the visualization of water transport in concrete [57]. The scattering strength of a particle depends not only on its shape and size, but also on its orientation relative to the incoming wave field. A fiber, for example, will scatter light strongly in the direction perpendicular to its long axis, but hardly along this direction. This fact can be exploited to detect the orientation of anisotropic scatterers in a sample, both in two and three dimensions [58], [59].

These are just some examples of application studies with X-ray grating interferometry. The last decade has seen an impressive development of the technique and efforts to push it to commercial applications are growing, e.g. with the integration of a grating interferometer into a commercial micro CT scanner [60]. A key for the success of the technique is the advancement in the fabrication of the gratings, which poses a challenge to microfabrication techniques due to the small period and large absorber thickness required for the gratings.

2.2.5 Grating requirements

Two types of gratings are used in a Talbot-Lau interferometer, a phase grating and two absorption gratings. While the necessary periods are in the same range for all gratings, the other properties differ a lot: While the absorption in the lines of an ideal phase grating is zero, the absorption in an ideal analyzer grating is 1, meaning no photons can pass through. Real gratings can of course reach neither of these ideal values.

Phase gratings

The period and phase shift of the phase grating define the overall length of the setup. The main requirements for a phase grating are

- Low photon absorption
- Well defined lamella height
- Excellent structure quality

Two factors contribute to the absorption: The lamellas and the substrate. Light materials are therefore favored for both, although the determining factor for the lamellas is not the absorption, but the ratio of the real to the imaginary part of the refractive index. A large real part means that a lower structure height is required to achieve a defined phase shift. The most common materials for phase grating lamellas are silicon and Nickel. For higher energies, one can also consider gold, because of its high refractive power for X-rays. The height of the lamellas defines the phase shift, and a good control over it ensures a high visibility over the whole field of view. As this parameter is hard to control and most manufacturing techniques have an inherent height gradient that cannot always be overcome, the height is usually specified to a target value +/- 10 %. Higher precision is often not needed because of the polychromatic spectrum of X-ray tube sources; if the grating design energy is slightly off the optimum, there is still sufficient flux at this energy.

Absorption Gratings

The absorption gratings are the biggest challenge for fabrication. The period is defined by the interference pattern created by the phase grating, and the height must be sufficient to absorb enough photons to achieve a high visibility in the stepping curve. Gold is the material of choice for the absorption lines because of its high density and high atomic number, which both lead to a relatively small penetration length for X-rays. Absorption of 90 % of the incoming photons is considered a sufficient value for an absorption grating; Figure 2.10 shows a plot of the necessary gold height to achieve this goal as a function of the photon energy. At around 40 keV, the necessary height for this absorption surpasses 100 µm, and together with the demand for grating periods of a few micrometers, the necessary aspect ratio (AR, defined as the structure height over the structure width) for the grating lines can be 100 or larger.

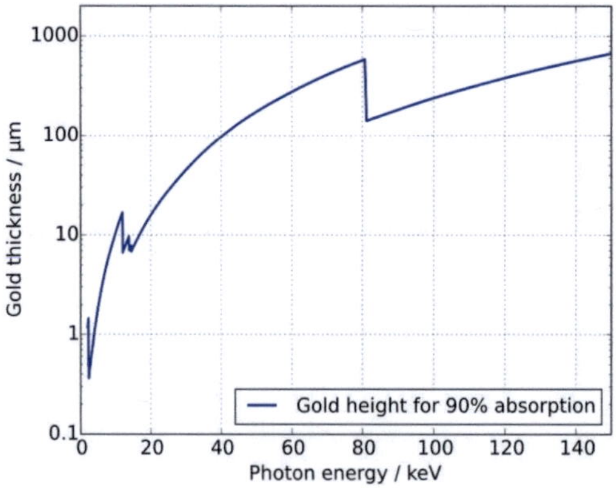

Figure 2.10: Plot of the necessary gold height for an absorption of 90 % of all photons as a function of photon energy, based on data from [61]

These extreme aspect ratios call for specialized fabrication methods, and a variety of techniques has been applied to fulfill the needs of X-ray grating interferometry. The first experiments were done using gratings fabricated by deep reactive ion etching (DRIE) [62], other groups used an imprinting process to structure metallic glass [63] or fabricated each grating lamella individually by a layer deposition technique [64]. Recently, a metal assisted wet etching technique has also gained attention, which is particularly suited for the fabrication of nanometric structures [65]. For micrometric structures with an extreme aspect ratio and a comparatively large area, the currently most powerful technique is the so-called X-ray LIGA process, a combination of X-ray lithography with electroplating, which is described in the following section.

3 X-ray lithography

The word lithography comes from the Greek words *lithos* (stone) and *graphein* (to write), and originally referred to a printing technique using an etched limestone as print master. Nowadays, any fabrication technique where a pattern of some sort is transferred to a (typically photoactive) polymer layer is called lithography. The most important lithography application certainly is UV lithography used in semiconductor chip fabrication, and the development of lithographic techniques has been the key enabler for the impressive advancement of microelectronics in the past decades. The scientific literature contains numerous examples for other lithography techniques like electron beam, two-photon or nanoimprint lithography [66]–[68], and the choice of technique greatly depends on the desired properties and dimensions of the structure to be fabricated. When it comes to microstructures with high aspect ratio, which means the ratio of the height of a structure to its width, X-ray lithography with subsequent electroplating is an excellent choice [13], [69], [70]. The process is also called "direct LIGA", which comes from the German acronym LIGA for "Lithographie, Galvanik und Abformung", meaning "lithography, electroplating and molding". The molding step is left out in the direct process, because demolding becomes difficult at high aspect ratios. An overview of the process can be seen in Figure 3.1.

The starting point is a substrate, with a conductive electroplating seed layer, which is coated with a photoresist, usually by spin-coating (Figure 3.1 A). The coated substrate is then exposed with soft X-rays using a mask with the grating pattern (B). If a negative photoresist is used, the irradiated sample undergoes the so-called post-exposure-bake, in which the exposed parts of the resist are crosslinked and become insoluble (C). The other parts of the resists are then washed away in a development step (D), leaving a polymer template which can then be filled with a metal like gold or nickel (E). Depending on the aspect ratio and intended use, the remaining photoresist can be removed by plasma etching (F) or left in the structures.

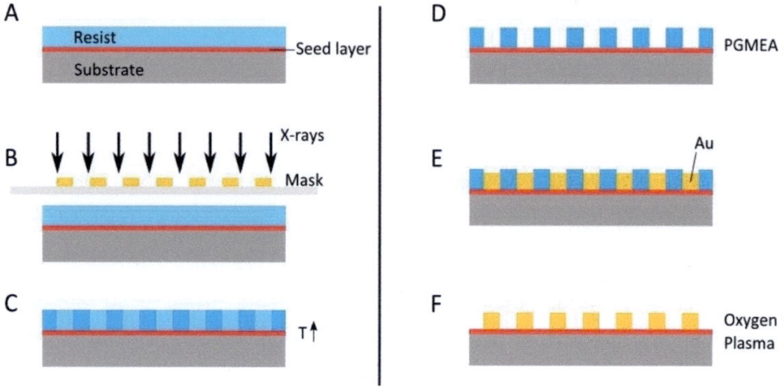

Figure 3.1: Schematic overview of the Direct LIGA process using a negative photoresist

The following sections describe the process in more detail as it is established at the institute of Microstructure Technology of the Karlsruhe Institute of Technology, for the resist mr-X, which is an epoxy-based negative resist. The properties of this photoresist are a main theme of this work; a dedicated chapter on this topic follows the process description.

3.1 Substrates

The choice of the substrate has several effects on both the fabrication process and the use of the final structure as X-ray grating. The fabrication process places some constraints on the substrate; it must have the following properties

- Mechanically stable
- Resistant against solvents used in the process
- Resistant against radiation damage
- Thermally stable in the temperature range -40 °C < T < 100 °C
- Conductive surface

The standard material for the bulk substrate is silicon, because it is comparatively inexpensive and easily available with excellent surface quality. As pure silicon is not a good conductor at the temperatures used in electroplating, it must be coated with a conductive material. Two different conductive layers are used in the process at the Institute of Microstructure Technology; titanium with an oxidized surface and a chromium base layer with gold on top. The oxidation of the surface of titanium increases the surface roughness and thus leads to a larger surface area which provides excellent adhesion of both photoresist and electroplated metal structures. On the other hand, it reduces the conductivity and makes height control in electroplating more difficult. For good height control it is favorable to use a Cr/Au coating. The effects of the conductive layer in the irradiation are explained in the following section.

The bulk material influences the overall X-ray transmission of the grating. While at high energies, the absorption in silicon can be neglected, it poses a problem for low energies, which is why alternative substrate materials have been investigated in this work.

3.2 Mask and exposure

The exposure in X-ray lithography is done with X-rays through a shadow mask that is fixed to the substrate by a specifically designed holder. Recently, another approach with a mask fixed to the synchrotron beam and a moving substrate was introduced [71]. The mask must fulfill two main criteria: High transparency in the openings and high absorption in the structures. These criteria resemble the criteria for the final absorption gratings, but for the exposure with so-called intermediate masks, a height in the order of 2 µm gold is sufficient to block enough intensity (see also Figure 2.10). This limits the aspect ratio of the mask structures and allows the use of direct laser writing for coarser structures down to about 5 µm period and electron beam lithography for smaller structures. The standard substrate for the masks is a 2.7 µm titanium foil, which is on a carrier wafer during mask fabrication and afterwards glued to an invar frame and lifted off the carrier. The stability of the membrane, particularly its transfer to a frame without waviness or bulges, limits the size of the mask; the technique is suited for the area of a 4-inch wafer but not easily upscalable.

In this work two beamlines of the ANKA synchrotron have been used for the exposure: LIGA I and LIGA II. Both beamlines are placed behind the same 1.5 T bending magnet and a mirror is used to cut off the high energy photons. The calculated spectra of the two beamlines are plotted in Figure 3.2. The main difference between the two beamlines is the energy range, the largest part of the flux at LIGA I is between 2.2 and 3.3 keV, while LIGA II has considerable flux up to about 12.5 keV. This means, that at LIGA I, an intermediate mask as described above is sufficient, but for exposure at LIGA II, the mask first must be copied to structures with higher aspect ratio, which is done with X-ray lithography at LIGA I. Masks fabricated like this are called working masks. Soft X-rays are attenuated and scattered easily, also when going through air. The exposure therefore takes place in a chamber that is first evacuated and then filled with helium to a pressure of about 100 mbar, to allow convectional heat transfer from the mask. As a very light element, helium barely influences the beam. The beam dimensions are similar in both beamlines, with a width of about 100 mm and a height of about 8 mm FWHM. For the exposure of a whole 4-inch wafer, it is moved up and down through the beam at a carefully controlled speed.

The choice of beamline for grating fabrication mainly depends on the height of the desired resist structures; up to about 100 µm, LIGA I is used, LIGA II is better suited for thicker resist layers. The dose deposited at the point where the X-rays enter the resist ("top dose") should not differ greatly from the dose deposited near the substrate ground ("bottom dose"). This is important, because the bottom dose must be high enough to completely crosslink the photoresist; but if the ratio of top to bottom dose is too high, the resist on the top of the structures will be overexposed, which can lead to closed structures that cannot be electroplated. In practice, the top/bottom ratio should be kept below 3, which is easier to achieve at LIGA II because the higher energy means less absorption and therefore a more homogeneous intensity over the resist height. Decreasing the top/bottom ratio at LIGA I is possible by introducing filters to harden the spectrum. Exposure doses are calculated using the program DoseSim, which takes into account the spectra of the beamlines, the filtering through mask materials and optional additional filters as well as the absorption in the photoresist [72].

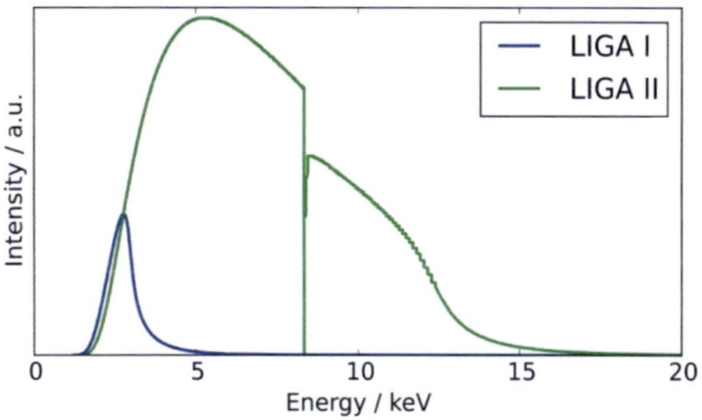

Figure 3.2: Unfiltered spectra of the beamlines LIGA I and II at the ANKA synchrotron

While the top/bottom ratio is better at high photon energy, it also has a disadvantage: In the soft X-ray range, the dominant process for photon absorption is the photoelectric effect (see section 2.1.3), where the photon energy is completely transferred to an electron. These electrons then travel on complicated paths with numerous interactions with other electrons; the mean length of such a path, and consequently the volume in which one initial photon leads to dose deposition, depends on the initial energy of the electron. This leads to a lower limit of the grating period that can be fabricated as a function of the photon energy. Currently, the smallest period fabricated at the LIGA II beamline is 4.8 µm, while at LIGA I a period of 2.4 µm is possible, smaller periods are under investigation at both beamlines.

The absorption in the photoresist is not the only process contributing to the dose deposited in it; the effects of the substrate must also be considered. A drawback of the titanium layer is that its Kα-line is at 4.95 keV, which can be excited by the illumination at the beamline LIGA II. As the fluorescence radiation is isotropic, it also exposes resist in areas under the mask absorber. If a chromium/gold layer is used, there is no fluorescence from the substrate, because the Cr layer is only a few nm thick and the Kα-line of the thicker gold layer is at 82 keV and can thus not be excited by the exposure. Both for Ti and Au, the X-ray absorption is a lot higher than for the photoresist, which means that more photoelectrons are created, that can diffuse into the photoresist from the substrate material, leading to additional dose deposition near the substrate ground. Because of its high atomic number, this effect is far larger for gold than for Ti. This additional dose can lead to crosslinking of resist also between the lamellas close to the substrate and prevent electroplating. As the layer is thin and often not fully crosslinked, it can be removed by reactive ion etching after development and prior to electroplating.

3.3 Development and drying

After exposure, the crosslinking of the photoresist takes place at elevated temperature, for a more detailed description see chapter 3.5. After the crosslinking reaction is done, the unexposed parts of the resist must be removed. This is done using the solvent propylene glycol methyl ether acetate (PGMEA), which is exchanged once after a time that is dependent on the structure height. The solvent used in the second step is saved to be reused in the first step for another sample to save solvent costs. After the second step, the structures are rinsed using Isopropanol. From the isopropanol bath, the structures must be dried. During drying, the Isopropanol evaporates gradually, and at the boundary between liquid, vapor and structures, capillary forces occur due to the surface tension of the liquid. The pressure p_a on the convex side of a curved liquid surface is always lower than the pressure p_i on the concave side. This relationship is described by the following equation [73]:

$$p_i = p_a + \frac{2\sigma}{r}. \tag{2.14}$$

Here, σ denotes the liquids surface tension and r the radius of curvature of the liquid surface. This pressure pulls the two sides of the capillary together, and because the liquid level on the two sides of one grating line will not be equal throughout the drying, it creates an asymmetric force that can be large enough to deform structures and even break them off the substrate. The surface tension is a function of the temperature and can be decreased with rising temperature. To make use of this, structures are usually dried in a convection oven at 35 °C, which reduces, but does not eliminate capillary forces. A way to eliminate them is freeze drying, a process originally developed for the preparation of biological samples. In this technique, the liquid is frozen and then sublimated at reduced pressure. In this way, the liquid phase is completely avoided and no capillary forces occur. As the freezing point of Isopropanol is very low (-89.5 °C), the solvent must be exchanged to avoid unnecessary thermal stress to the sample. A good liquid for freeze drying is Cyclohexane, as its triple point lies only slightly below ambient conditions (5319.6 Pa, 6.82 °C [74]) and it is fully miscible with Isopropanol. After the solvent exchange, the liquid is frozen inside a chamber which is then evacuated to speed up the sublimation process. After complete sublimation, the sample is slowly heated up to room temperature and then taken out of the vacuum. It has been shown that the maximum possible aspect ratio of X-ray lithography structures can be increased by a factor of 2.25 using this process instead of conventional drying [75].

3.4 Electroplating

The electroplating of Nickel structures for phase gratings is done in a specially developed bath [76]. Provided the starting ground is free from residuals and the surface area is precisely known, it provides excellent height uniformity with a deviation of +/- 200 nm. As a small variation of the duty cycle in the polymer grating matrix can change the electroplating area significantly, the actual precision of the absolute height value is less than this, but still sufficient for the application as phase gratings. Moreover, it is possible to achieve relatively smooth surfaces, which prevents unwanted scattering.

Electroplating of gold for absorbing gratings is done using a sulfite-based electroplating bath. It provides good adhesion of the metal structures, is well compatible with the polymer matrix and yields structures with low inner stress [77].The temperature of the electroplating bath is 55 °C, which can lead to problems for two main reasons; the crosslinking reaction during the post bake can create inner stress in the polymer matrix, that can be released at higher temperatures and lead to a deformation of the matrix. Moreover, the coefficient of thermal expansion of the substrate material and the photoresist matrix differ by an order of magnitude when using silicon substrates, creating thermally induced stress during heating. For these reasons, a lower temperature electroplating bath operating at 30 °C is under test, which shows good results but at a lower deposition rate than the standard bath at 55 °C [78].

3.5 Photoresist in X-ray lithography

Nowadays, many photoresists are available for different lithographic techniques. Independent of application, they can be grouped in two major categories, the positive and negative photoresists. For positive resists, the parts that are exposed with radiation are dissolved, which for mask-based lithography techniques results in a geometry identical to the absorber structures on the mask. In negative resists, a crosslinking reaction is triggered by irradiation, and the unexposed parts are washed away in development, resulting in the pattern being the negative of the mask absorber pattern. In X-ray lithography negative tone resists are based on the well-known resist SU-8, originally developed for UV lithography [79]. Owing to their chemical amplification, these resists require significantly less exposure dose than PMMA, which is a commonly used positive resist.

The two main parameters to describe the performance of a photoresist are sensitivity and contrast. For a negative tone resist, the sensitivity is related to the minimum dose needed to crosslink the material. A higher sensitivity means a lower necessary exposure dose, but at the same time it increases the sensitivity to secondary effects, such as residual dose deposited under the mask absorbers in X-ray lithography. A highly sensitive resist can therefore limit the achievable resolution. The contrast γ is defined as the logarithm of ratio of the dose D_2 necessary to crosslink 90 % of the resist material and the dose D_1 necessary for 10 % crosslinking, which can also be used to quantify the sensitivity [80].

$$\gamma = log_{10}\left(\frac{D_2}{D_1}\right)^{-1}$$

In this work, the production of X-ray diffraction gratings with the resist mr-X, developed by the company *microresist technology*, Berlin, is investigated. The following sections describe its composition and function.

3.5.1 Composition of mr-X

The principal components of a negative tone photoresist are resin, photoactive compound, and solvent. The basis for mr-X is the SU-8 resin with the photo acid generator (PAG) Triarylsulfonium/hexafluoroantimonate salt and Gamma-butyrolactone (GBL) as solvent [81]; the structural formulae for these components are shown in Figure 3.3. The SU-8 resin monomer consists of 4 interlinked units of bisphenol-A (BPA) which is the basis of many epoxy resins. The two hydroxyl groups of the BPA units are replaced with epoxy groups, providing in total 8 epoxy groups per monomer in the ideal case (hence the 8 in the name). In practice, the synthesis of the monomers does not yield a perfect epoxidation, and some variations between different batches of the resin are to be expected [82]. The PAG is an onium salt, as it is used in several chemically amplified resists, and irradiation triggers its dissociation. The solvent GBL was also used in the original SU-8 formulation [15], it is fully miscible with water and most common organic solvents.

Mr-X comes in two variations, called mr-X 10 and 50, which have different solvent content and therefore different viscosity. The low viscosity variation, mr-X 10, has the higher solvent content and can be used for the preparation of layers between 5 µm and 100 µm, while the higher viscosity mr-X 50 is used for layers with more than 50 µm thickness.

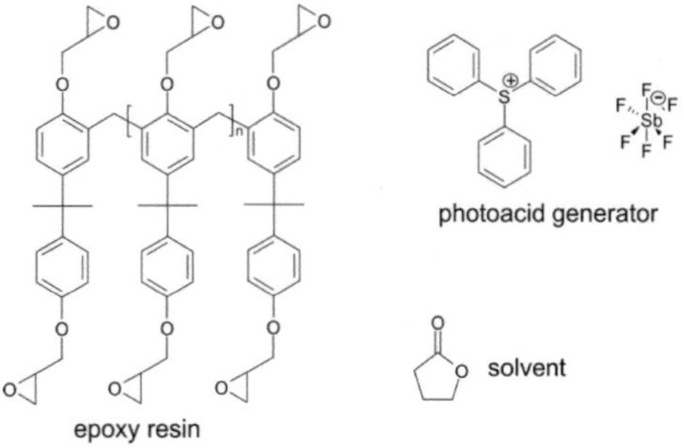

Figure 3.3: Structural formulas of the main components of the mr-X photoresist, adapted from [83]

In addition to the basic components resin, PAG and solvent, mr-X contains an additive with lower molecular weight than the resin monomers to increase the adhesion to substrates and a buffer, whose function is described in the post-exposure-bake section.

3.5.2 Layer preparation

The preparation of mr-X layers is done using a spin-coating process. The resist is placed on the substrate and through fast rotation it is evenly distributed over the substrate. The final thickness of the layer depends on the rotation speed. In the process, and mainly in the last phase, solvent already evaporates from the resist [84]. After the coating, the resist is soft-baked at 95 °C on a hot plate to reduce the solvent content further. Three main problems arise when the solvent content is too high

- Formation of bubbles during crosslinking reaction
- Reduced mechanical stability of final structures [85]
- Reduced contrast due to better diffusion of protons [85]

On the other hand, a solvent content that is too low may lead to a lower degree of crosslinking in the final structure, because of the reduced diffusion of protons. For SU-8, reported values for acceptable residual solvent content vary around 7 %, while in X-ray lithography experiments with mr-X, it was found that the solvent content should be below 3 % for good results in structuring [86], [87].

3.5.3 Post exposure bake

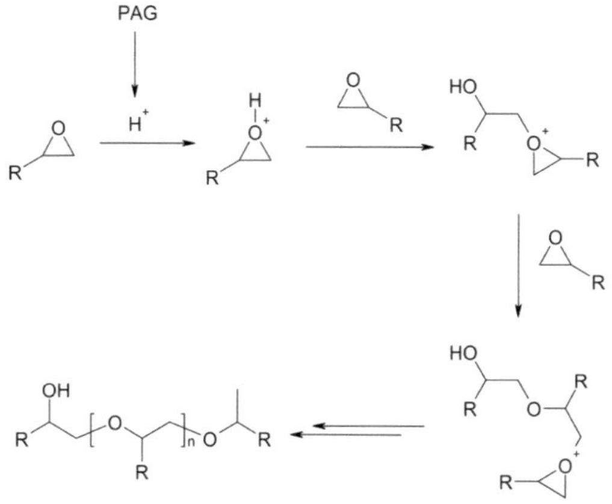

Figure 3.4: Crosslinking mechanism of the epoxy groups in the mr-X photoresist, adapted from [83].

Figure 3.4 shows an illustration of the crosslinking mechanism in the photoresist. When absorbing a UV photon, the PAG dissociates and $HSbF_6$ is created which is a strong Lewis acid and thus acts as proton donator. The proton associates with the oxygen atom of an epoxy ring at an oligomer of the resin and induces the opening of another ring which then binds to the first via a methylene bridge. The proton is not consumed in this reaction, and can thus trigger further ring openings. The absorption of soft X-rays is mainly due to the photoelectric effect, a photon liberates an electron which creates a cascade of secondary electrons. These secondary electrons lead to dissociation of the PAG via impact ionization, and ultimately trigger the same chemical amplification as UV irradiation, the polymerization mechanism is therefore the same for both techniques [88], [89].

The crosslinking reaction starts immediately upon irradiation, but due to the limited mobility of the protons only creates a very small degree of crosslinking. In thicker layers, a latent image of the final structures can already be seen after irradiation. To achieve a high degree of crosslinking, the sample has to undergo a post-exposure-bake (PEB). This step takes place at an elevated temperature, e.g. 75 °C. During the PEB, the molecular weight increases exponentially. The polymerization eventually stops if steric effects hinder the binding of two epoxy sites, or the proton encounters a quencher, such as the buffer that is added to the mr-X. The buffer limits the space, in which a single initial proton can trigger polymerization. This has two effects on the resist – it lowers the sensitivity, i.e. it increases the dose necessary to crosslink the material, and it increases the contrast, making the resist less vulnerable to dose deposition in areas that are not primarily irradiated [81].

Due to the increase in molecular weight in the crosslinking, the photoresist becomes insoluble in most solvents by the PEB. The unexposed areas can then be developed; this is done using PGMEA, as described in section 3.3. The necessary development time can be long, as overdevelopment is not an issue, contrary to the case of positive resists like PMMA.

A very important side effect of the crosslinking is that it reduces the space in between the resin molecules, and thus the overall volume of the resist shrinks. For SU-8, a shrinkage of about 7 % has been reported in a study that looked at the height of crosslinked versus uncrosslinked resist [90]. In the first phase of shrinking, the volume is just decreasing, and the material still shows viscous-plastic behavior, the shrinkage introduces no stress. Once a certain degree of crosslinking is achieved, the material shifts to a rigid elastic phase. This transition is called the gel point of a polymerization; it corresponds to individual polymerization centers interconnecting [91]. After this point, the shrinkage in one polymerization center affects neighboring areas, leading to stress in the structures. This stress can lead to deformation of structures and can be so severe that it causes cracks. The investigation of the crosslinking reaction and its effects on volumetric shrinkage of the resist is another main theme of this work, details are described in chapter 4.4.

4 Photoresist Characterization

The thermomechanical properties of the photoresist are a major determining factor for the quality of lithographic structures. This chapter deals with the determination of properties like the Young's modulus, tensile strength, volumetric shrinkage and thermal behavior of both the unexposed and exposed resist. The influence of process parameters, namely the exposure dose, post-bake temperature and substrate storing time were varied to assess their influence. Unless otherwise stated, all results were obtained using the photoresist mr-X 50. It should be noted that some variations in the yield and quality were seen between gratings fabricated with resist from different bottles that were bought from the manufacturer mrt. When a new bottle is used, contrast and sensitivity are evaluated by studying the so-called contrast curve [92]. These samples are named with the letter K for the German "Kontrastkurve" and a consecutive number, e.g. K110. This number is then also used to identify the resist bottle. Precautions were taken to always use resist from the same bottle in all experiments that were designed to assess the influence of a single process parameter on the properties of the final structures. In some experiments, this practice had to be abandoned because the respective resist bottle was emptied in the meantime. These cases are explicitly mentioned in the description.

4.1 Standard parameters

The fabrication of gratings with the resist mr-X has been the subject of several research projects at IMT already, among them the INNOLIGA project, and the dissertations of Johannes Kenntner and Jan Meiser [87], [93], [94]. From these projects, a set of standard parameters emerged for the different process steps, which was used as a basis for the variations performed in this work. The most important parameters for the main steps are listed below; every time a standard parameter is mentioned in the experiment descriptions, it is a reference to the parameters listed here.

Soft bake

All spin-coated substrates are soft baked immediately after the layer deposition on a hot plate. For a target thickness below 40 µm, the substrates are placed directly on the hotplate with the soft bake temperature of 95 °C. The holding times are dependent on the thickness: 0-12 µm: 10min; 13-25 µm: 15min; 26-40 µm: 20min. If the target thickness exceeds 40 µm, a stepwise soft bake is employed, where the sample is heated up to 75 °C in 30 minutes, held at that temperature for one hour, and then heated up to 95 °C in 30 min. This temperature is held for 4 hours, cooling down to room temperature also occurs gradually in 4 hours.

Exposure

The standard exposure dose varies with the beamline that is used for exposure. At LIGA I, the standard bottom dose is 140 J/cm^3, whereas at LIGA II, it is only 60 J/cm^3. At both beamlines, filtering is used to keep the top/bottom dose ratio below 3. Between mask and resist, a Kapton foil of 7.5 µm thickness is used to capture photoelectrons from the mask membrane.

Post exposure bake

For post exposure bake, the exposed substrates are placed inside a Heraeus Vacutherm oven. The oven is evacuated to a residual pressure below 20 mbar and upon reaching this pressure, all valves are closed and the pump is shut off. Then the temperature is increased to 75 °C in 20 min and held for 2 hours. Cooling down takes place overnight, and the oven is ventilated at the next morning. After the PEB, the substrate is stored for at least 24 hours until further processing.

Development and drying

The development time depends on the structure height, substrates with a height up to 15 µm are developed for 10 min in reused PGMEA, then 10 min in fresh PGMEA and afterwards rinsed for 10 min in Isopropanol. The respective times for higher structures are 30 min, 60 min, 20 min. Drying takes place either in a convection oven at 30 °C or via freeze drying after a solvent exchange to cyclohexane using a Labconco FreeZone Tray dryer. For a more detailed description of the freeze drying process, see [95].

4.2 Thermal analysis by differential scanning calorimetry

Differential scanning calorimetry (DSC) is a technique for thermal analysis of a material. The measurement signal is the difference in the amount of heat needed to increase the temperature of a sample in a container compared to an empty container of the same material and weight. It is a sensitive tool for detecting phase transitions and chemical reactions happening at different temperatures inside a material. All measurements reported here were carried out using a Netzsch *DSC 204 Phoenix* device. The samples were placed inside aluminum cups; one sample had an average weight of 15 mg.

The purpose of these measurements was to establish the temperature range, in which other experiments had to be carried out, especially concerning the PEB. For this, three different samples were prepared in the same way up to different stages of the process. All samples were prepared on a silicon wafer with an oxidized titanium layer. Before spin-coating the resist, the substrates were cleaned using oxygen plasma. Then, a layer of mr-X 50 was spun on to get a layer of ca. 100 µm, and the standard soft-bake parameters were used (see section 4.1). The first sample was prepared from a wafer after the soft bake; it was not exposed to X-ray or UV radiation. The second sample was exposed at LIGA II with a bottom dose of 60 J/cm^3 without any mask or filtering, but did not undergo the PEB; the third sample was exposed in the same way and subjected to the standard PEB at 75 °C. The results from the DSC measurements are plotted in Figure 4.1 A. The temperature was scanned from 0 °C to 400 °C with a heating rate of 10 °C/min.

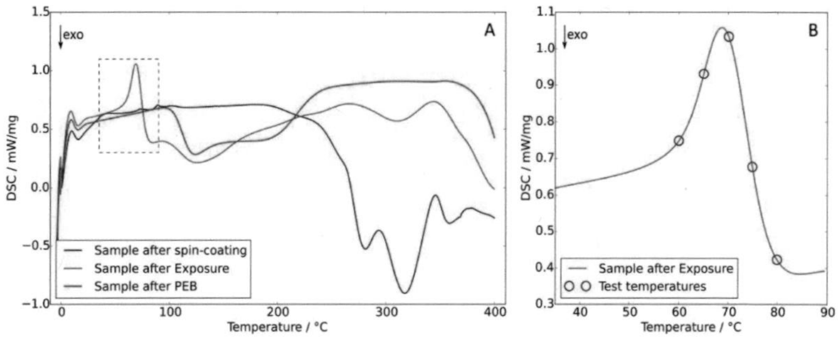

Figure 4.1: **A**: DSC curves of one sample after spin-coating and softbake, one after X-ray exposure and one after standard PEB at 75 °C; **B**: Close-up of dashed area of sample after X-ray exposure with temperatures for PEB test measurements

The blue curve in Figure 4.1 A shows the sample that was measured after softbake and before exposure. An interesting feature is that the curve shows a change in slope at around 40 °C, which is a hint to a glass transition in the material. In measurements with SU-8, this transition was found to be around 50 °C [96]. However, a disagreement of this magnitude can easily be caused by the sample preparation, especially by variation in the solvent content. At around 200 °C, an exothermal process starts, which can be identified with thermal crosslinking of the epoxy resin. At this point, a crosslinking starts even without the influence of the photoinitiator. The green curve shows the sample after X-ray exposure, the most important sample for the determination of parameters for other experiments; the transition at 40 °C is not visible here, which is an indication for a small degree of crosslinking already being present after exposure. Between 50 °C and 60 °C, there is the onset of a peak that signifies the start of the crosslinking reaction, this peak is magnified in Figure 4.1 B. It is endothermic up to a temperature of ca. 70 °C and then becomes exothermic. The standard PEB temperature of 75 °C lies in the exothermic part of the curve. From the analysis of the reaction footprint in the DSC, a useful starting point for PEB temperature variation is 60 °C. Below this temperature, no significant change in the material is to be expected. The temperatures used in further experiments are indicated with blue circles in Figure 4.1 B.

The red curve in Figure 4.1 A shows the curve of the sample after the standard PEB. The peak of the reaction seen in the green curve is gone here, which is expected because the reaction has already occurred in this sample. At temperatures over 90 °C, the reaction starts again, leading to a further crosslinking. This behavior is also expected, as it is known that a so-called hardbake can increase the degree of crosslinking and the stiffness of SU-8 structures [97].

4.3 Mechanical analysis by tensile testing

Young's modulus, ultimate tensile strength and elongation at break are the key properties that describe the mechanical stability of a specific material. For any simulation of structural deformation of lithographic structures, these values must be known as input parameters. All these properties can be influenced by the structure dimensions and fabrication method, and thus must be tested on microstructures with dimensions comparable to the grating structures. This was done using a micro tensile testing setup developed for high temperature testing of metal alloys at the Institute of Applied Materials at KIT [98], [99], based on earlier work by Zupan et al. [100]. The schematic of the setup is shown in Figure 4.2; the dog-bone shaped sample is mounted on a fixed support at the top and the bottom end is fixed to a holder that is connected to a step motor and force measurement device with a wire. The step motor is used to exert a continuously increasing uniaxial tensile load on the sample, and the deformation of the sample under this load is recorded by a camera. As the force is increased, a LabVIEW script simultaneously records camera images and force measurement values at an adjustable time interval, which was chosen as one frame per second.

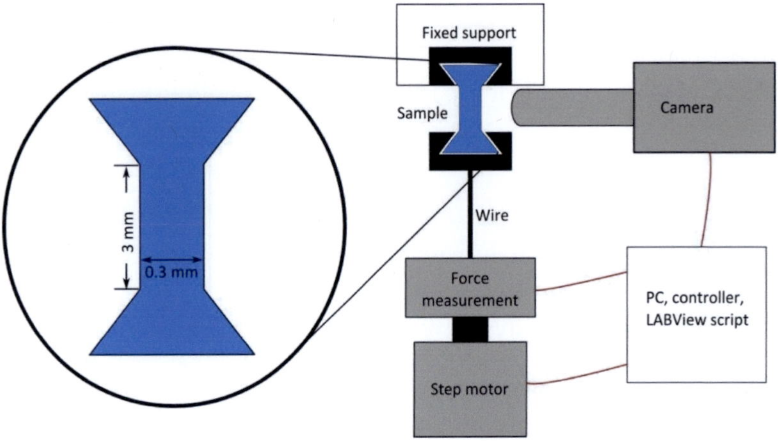

Figure 4.2: Schematic of the tensile testing setup used with sample dimensions

The stress exerted on the sample was calculated using the known dimensions of the sample with the data from the force measurement, and the strain was calculated using an image correlation algorithm implemented in a Matlab script that tracks points on the camera image from frame to frame, yielding the displacement of each point as a function of the image [101]. Combining these values yields the stress-strain curve of the material, which reveals many details about the material behavior. An exemplary curve for the photoresist samples measured is shown in Figure 4.3.

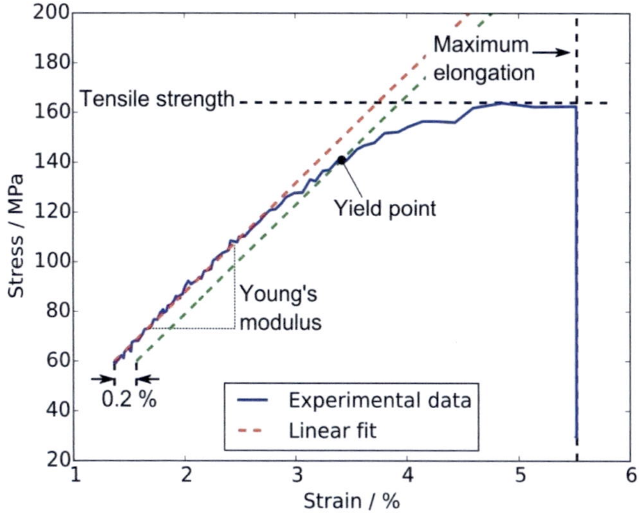

Figure 4.3: Exemplary stress-strain plot with the significance of the extracted data

The experimental data shown in blue in Figure 4.3 show a linear relation of stress and strain at low strain; the curve then flattens at higher strain until the sample breaks at its point of maximum elongation. The linear region of the curve is the domain of elastic deformation, where the sample returns to its original shape upon release of the stress. The slope of this linear regime is defined as the Young's modulus E, also called elastic modulus. Materials with a high Young's modulus experience little deformation with stress; it is thus a quantification of the stiffness of a material. There are different ways to define the end of the elastic regime; a commonly used one is illustrated in Figure 4.3, where a line is drawn parallel to the fit of the experimental data that is shifted by 0.2 % to higher strain. The point at which this line intersects the stress-strain curve is called the yield point of the material and marks the transition to the regime of plastic deformation, where an initial deformation of the sample induces a permanent deformation and the sample does not return to its original shape upon release of stress. Materials that show little to no plastic deformation are called brittle, e.g. glass; those that show a large plastic deformation are called ductile, aluminum is an example of a ductile metal. Note that the yield point defined in this way is only an orientation — the point where actual plastic deformation starts can be different depending on the material.

The ultimate tensile strength (UTS) of a material is the highest stress it withstands without rupture and is easily identified in the stress-strain plot as the highest stress value; the highest strain value is the elongation at break of the sample.

The Young's modulus for mr-X structures fabricated by X-ray lithography has been measured by Lemke looking at the deflection of a beam when pressing on it with a defined force [81]; this method measures a different property than the uniaxial tensile test used here, because the deformation includes torsional stress, to which a material may react differently. Nevertheless, the values provide a reference for the tensile tests and were found to be highly dependent on the process parameters, with values ranging from 2.3 GPa to 5.1 GPa. The samples for these measurements were soft baked at a temperature of 95 °C and the PEB was done at 65 °C. The variation was seen to be due to the exposure dose and the residual solvent content, which are thus also expected to influence the Young's modulus measured by tensile testing. The values reported by Lemke are in the same range as measured for SU-8, where it was found that the Young's modulus depends on the degree of crosslinking and can be increased by performing a hardbake at 200 °C [97], [102].

4.3.1 Variation of exposure parameters

As the exposure dose was found to have a major impact on the Young's modulus in earlier works by Lemke [81], experiments were designed to assess the influence of this process parameter. As the tensile test specimens must be lifted off the substrate for the measurement, all samples for these measurements were prepared with a thickness between 100 µm and 125 µm on Si wafers with a titanium layer that was not oxidized. After development, the Ti layer was etched in HF until the specimens could easily be removed. Due to the excellent chemical stability of the crosslinked photoresist, this is not expected to influence the mechanical properties significantly, and no deformation or dissolution of the specimens due to the etching could be observed.

Samples were exposed at both LIGA I and LIGA II, and the bottom dose and filters were varied. The parameters are summarized in Table 1. One sample set contains 70 tensile test specimens, of which 15-25 were analyzed; all reported values are the average over these specimens, with the standard deviation given as error. The bottom dose at LIGA I was varied from 105 J/cm^3 to 210 J/cm^3, always using 125 µm of Kapton as Filter, which results in a top/bottom ratio of about 3, depending on the actual sample thickness. At LIGA II, the dose was varied from 60 J/cm^3 to 120 J/cm^3, and two filter sets were used. The same Kapton thickness as at LIGA I resulted in a top/bottom ratio of 1.4. To simulate the influence of a working mask on a silicon substrate, another filter set was used here, with a nickel filter of 27.5 µm thickness and a 100 µm silicon filter. This filter set had previously shown good results in the structuring of thick mr-X layers. This reduces the top/bottom ratio to 1.1 and increases the exposure time roughly by a factor of 50, but the total exposure time was still below 20 min. For the standard exposure doses, 140 J/cm^3 at LIGA I and 60 J/cm^3 at LIGA II, two sets of samples had already been measured before by Amberger (samples 0442-5 and 0562-5) [78], and the experiments were repeated to assess the reproducibility of the results.

From the listing of the tensile testing results in Table 2, no clear dependence of Young's modulus, UTS, elongation at break and yield point can be identified. Instead, a plot of the results as function of the sample age at exposure (Figure 4.4) shows a trend towards both higher Young's modulus and UTS with increasing sample age, which saturates at a final level at a sample age of ca. 3 weeks. The difference in both values is considerable, the Young's modulus increases from the lowest value of 2.8 GPa to an average of 4 GPa, and the UTS increases from around 100 MPa to around 150 MPa. Both the yield point and the elongation at break do not seem to be influenced significantly.

Table 1: Parameters for tensile sample sets with varying exposure dose and spectrum

Sample number	Beamline	Filter	Bottom dose / J/cm³	Sample age at exposure / days
0602-5	LIGA I	125 µm + 7.5 µm Kapton	105	18
0529-5	LIGA I	125 µm + 7.5 µm Kapton	140	390
0442-5	LIGA I	125 µm + 7.5 µm Kapton	140	2
0601-5	LIGA I	125 µm + 7.5 µm Kapton	210	16
0562-5	LIGA II	125 µm + 7.5 µm Kapton	60	4
0530-5	LIGA II	125 µm + 7.5 µm Kapton	60	390
0566-5	LIGA II	100 µm Si + 27.5 µm Ni + 7.5 µm Kapton	60	120
0563-5	LIGA II	125 µm + 7.5 µm Kapton	120	4
0564-5	LIGA II	100 µm Si + 27.5 µm Ni + 7.5 µm Kapton	120	270

Table 2: Sample sets for tensile testing, ordered by sample age at exposure

Sample set number	Sample age at exposure	Young's modulus / GPa	UTS / MPa	Elongation at break / %	Yield point / %
0442-5	2	2.8 +/- 0.5	105 +/- 11	9.7 +/- 7.9	-
0562-5	4	3.3 +/- 0.2	99 +/- 3	4.6 +/- 0.4	-
0563-5	4	3.6 +/- 0.2	118 +/- 1	5.6 +/- 0.6	3.0 +/- 0.2
0601-5	16	3.9 +/- 0.2	145 +/- 9	6.5 +/- 1.1	3.3 +/- 0.3
0602-5	18	3.7 +/- 0.2	132 +/- 11	4.8 +/- 0.9	3.3 +/- 0.2
0566-5	120	4.1 +/- 0.3	139 +/- 8	4.3 +/- 0.7	3.3 +/- 0.2
0529-5	390	4.0 +/- 0.2	153 +/- 10	5.6 +/- 0.8	3.4 +/- 0.2
0530-5	390	4.1 +/- 0.2	150 +/- 7	4.9 +/- 0.7	3.4 +/- 0.3

With these results, it is obvious that the influence of the exposure dose can only be investigated by the comparison of samples with similar waiting time between spin-coating and exposure. This is the case for the samples 0562-5 and 0563-5, both with a waiting time of 4 days. The sample 0562-5 was exposed with the standard bottom dose of 60 J/cm³ at LIGA II with only Kapton filtering, the sample 0563-5 was exposed with the same spectrum and twice the dose, 120 J/cm³.The Young's modulus of the latter was measured to be 3.6 GPa compared to 3.3 GPa of the sample with the lower dose, an increase of ca. 10 %. The UTS increases from 99 MPa to 118 MPa with the doubling of the dose, and also the elongation at break rises from 4.6 % to 5.6 %. The situation is similar for a second set which can be used to track the influence of the exposure dose, this time at LIGA I: Sample 0601-5 was exposed 16 days after coating with a bottom dose of 210 J/cm^3, sample 0602-5 had a similar waiting time of 18 days and a bottom dose of 105 J/cm^3. With the increase in bottom dose, the Young's modulus increases from 3.7 GPa to 3.9 GPa, the UTS rises from 132 MPa to 145 MPa and the elongation at break shows an increase from 4.8 % to 6.5 %. From these two sample sets, it can be concluded that an increase in dose does enhances the mechanical properties of the resist structures in a similar way a hardbake does [97], it can thus be assumed that the higher dose results in a higher degree of crosslinking. The influence of the dose on the Young's modulus found here is not as large as seen by Lemke, which can be explained by the dose range in which the measurements were done. The values for the Young's modulus of around 4 GPa after sufficient waiting time match those reported by Lemke.

Another result from these measurements is that there is no significant influence of the exposure dose on the yield point, all samples except for the one with the shortest waiting time between coating and exposure agree within their standard deviation.

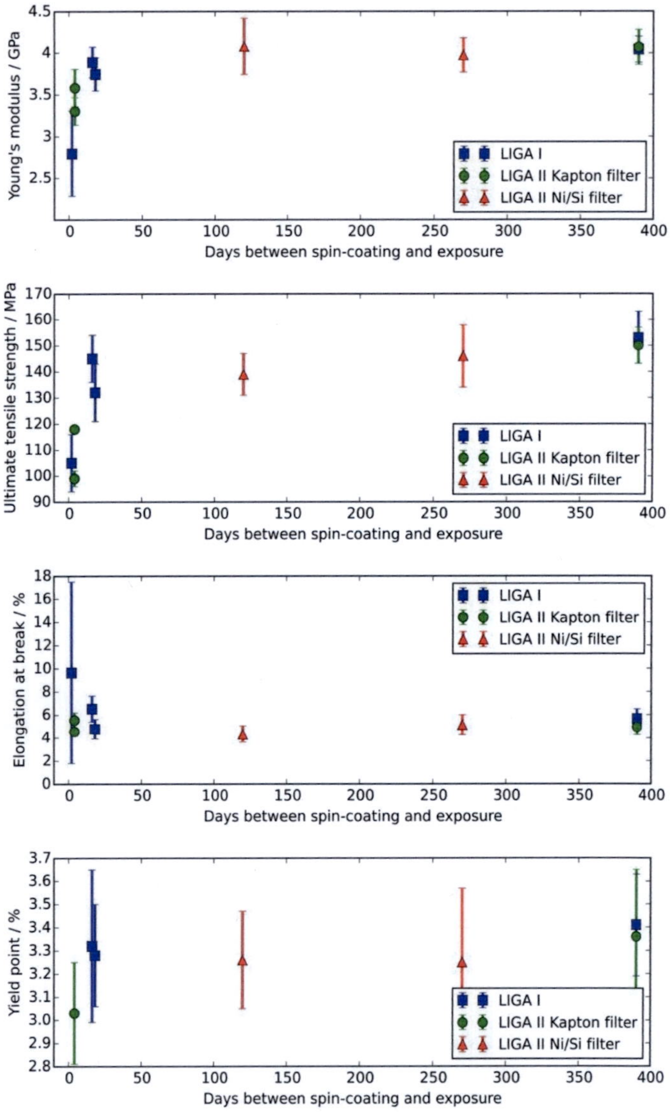

Figure 4.4: Results from tensile tests with variation of the exposure parameters as function of time between coating and exposure

The reason for the dependence of the mechanical properties on the substrate age is most likely to be found in the distribution of the residual solvent after the softbake. It is known that the solvent GBL can copolymerize with the SU-8 epoxy resin; excess solvent was found to act as a plasticizer [103]. It is also known that after the softbake, the solvent distribution over the height of the resist is not homogeneous, instead there is a solvent depletion region near the surface of the sample and the solvent content gradually increases with depth in the resist layer[104]. This gradient is expected to level out over time, yielding more homogeneous mechanical properties. A larger solvent content on the lower part of the tensile specimen and thus a higher plasticity would lead to a situation where the force exerted by the step motor in the tensile testing machine is countered mainly by the more rigid upper part. This results in increased stress in this part compared to the calculated stress and thus in a rupture at a smaller overall force.

4.3.2 Variation of storage time and temperature

To further investigate the influence of the substrate age at exposure, seven wafers were spin-coated on the same day with the same parameters and underwent the same softbake. Three of these samples were then stored at 40 °C on a hot plate after the softbake, the other four samples were stored in a climate controlled chamber, just as the samples of the dose variation experiment. The samples were then exposed at different times at LIGA II with the same bottom dose of 60 J/cm^3 and the same filter set using 100 µm Si and 27.5 µm Ni with 7.5 µm Kapton, to achieve a small top/bottom dose ratio and therefore a homogeneous material. They underwent the same PEB in the vacuum oven, the same waiting time between PEB and development and also the same development times. The tensile tests were carried out within 3 days after development for all samples.

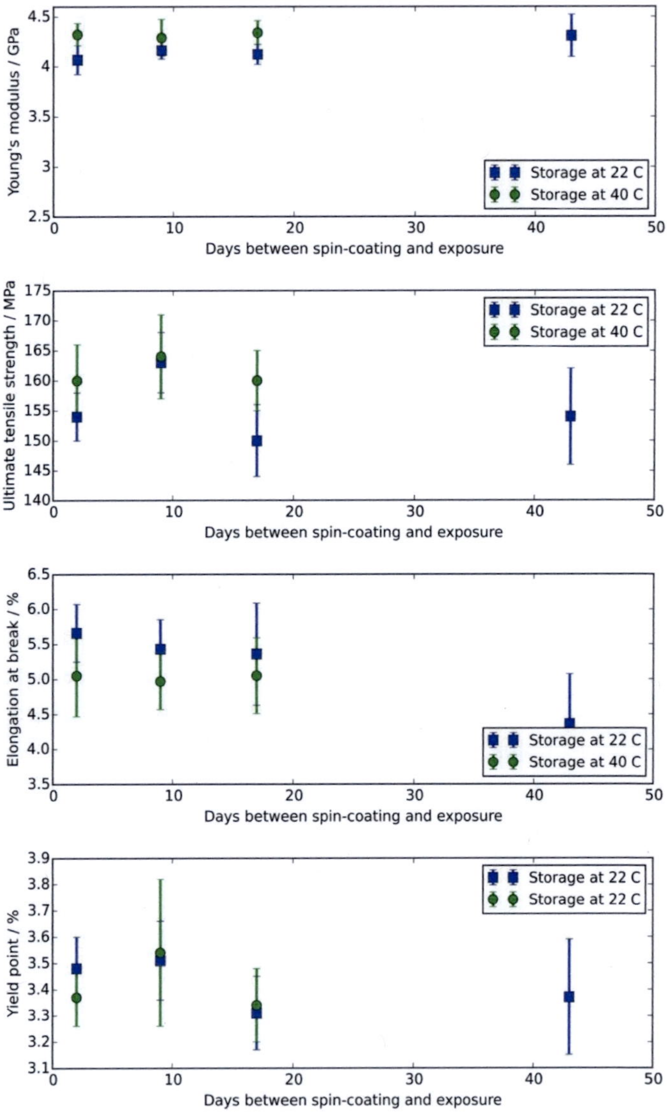

Figure 4.5: Results from tensile tests with variation of time between coating and exposure

Table 3: Sample sets and results from tensile tests with controlled substrate age and storing temperature with reference sample

Sample number	Sample age at exposure	Storage tempe-rature	Young's modulus / GPa	UTS / MPa	Elongation at break / %	Yield point / %
1004-5	2	22 °C	4.1 +/- 0.1	154 +/- 4	5.7 +/- 0.4	3.5 +/- 0.1
1010-5	2	40 °C	4.3 +/- 0.1	160 +/- 6	5.1 +/- 0.6	3.4 +/- 0.1
1006-5	9	40 °C	4.3 +/- 0.2	164 +/- 7	5.0 +/- 0.4	3.5 +/- 0.3
1008-5	9	22 °C	4.2 +/- 0.1	163 +/- 5	5.4 +/- 0.4	3.5 +/- 0.2
1005-5	17	22 °C	4.1 +/- 0.1	150 +/- 6	5.4 +/- 0.7	3.3 +/- 0.1
1007-5	17	40 °C	4.3 +/- 0.1	160 +/- 5	5.1 +/- 0.5	3.3 +/- 0.1
1009-5	43	22 °C	4.3 +/- 0.2	154 +/- 8	4.4 +/- 0.7	3.4 +/- 0.2
0566-5	120	22 °C	4.1 +/- 0.3	139 +/- 8	4.3 +/- 0.7	3.3 +/- 0.2

The results from the sample sets with controlled sample age at exposure and storing temperature are listed in Table 3 and plotted in Figure 4.5. Again, 15-25 tensile specimens were analyzed for each sample set, and the errors given in Table 3 represent the standard deviation. The results were expected to fall in the range of sample 0566-5 from the previous set of measurements, which had the same exposure parameters. It was found that the Young's modulus and elongation at break in the new series resembled sample 0566-5, but the UTS was consistently higher, and the yield point also showed a small increase, which means that the new sample set appears overall a little more brittle than the previous one. These differences can be explained by the fact that the samples originated from two different resist fabrication batches, in between which the resin batch had to be switched. The resin is the main factor determining the achievable mechanical properties; unfortunately, it can only be fabricated with a certain variance in the epoxidization, i.e. the number of epoxy groups per resin monomer. Different batches can therefore yield slightly different mechanical properties due to a difference in the number of sites per monomer available for crosslinking [83]. The increase in brittleness suggests a higher degree in epoxidization in the new batch.

The results within the series are more homogeneous than within the first series. One thing that can be concluded is that the samples with storage at 40 °C show no dependence of the mechanical properties on the substrate age at exposure. Except for the yield point of sample 1006-5, all experimental values obtained for Young's modulus, UTS and elongation at break lie no further apart than one standard deviation. This excellent agreement between the results of samples at different storage times suggests that storage at an elevated temperature of 40 °C for 24 h is enough to provide the same effect as the sample ageing of more than two weeks. However, the difference in mechanical properties between samples of different age was much less pronounced in this series compared to the previous series with variation in the exposure; the Young's modulus of the 'youngest' sample set with storage at 22 °C was found to be 4.1 +/- 0.1 GPa, compared to 4.3 +/- 0.2 GPa for the sample set with the longest storage time of 43 days.

Apart from the higher Young's modulus of the samples with elevated storage temperature, they also show less plastic deformation than the samples stored at 22 °C with the same age, which results in lower values of the elongation at break. With no visible influence on the yield point, this means that the heat-treated samples are slightly more brittle than the samples stored at 22 °C, which hints to a lower overall solvent content. The same is true for the sample with the longest storage time. In conclusion, no difference in the mechanical properties could be found between sample sets with a storage time of at least 24 h at 40 °C and a sample set that was stored for 43 days in a climate controlled cabinet at 22 °C, and the samples with either a storage at 40 °C or a long waiting time between coating and exposure show mechanical properties that are explained with a higher degree of crosslinking compared to 'fresh' samples.

4.4 Measurement of the volumetric shrinkage during crosslinking

Apart from the mechanical properties of the crosslinked resist, a major factor for the achievable structure quality is the volumetric shrinkage the resist undergoes during the PEB, as described in section 3.5.3. The measurement of cure shrinkage is a problem not only relevant to photolithographic techniques, but also in many other fields like microelectronics packaging or dental medicine. It can be measured indirectly via the deformation of a substrate with a resist layer on top [81], but this technique relies on exact knowledge of the elastic modulus and layer thickness of both the resist and the substrate. An approximate method is the determination of dimensional changes of structures during the PEB, such as the height of the structures that was used in reference [90], or lateral dimensions, which are more difficult to measure before and after PEB. However, both approaches are not direct volumetric measurements, but only measure one or two spatial dimensions. Various true volumetric methods are in use, like dilatometers, gravimetric methods or rheometric methods, most of which require relatively large samples and are difficult to operate with photosensitive polymers [105]. When it comes to the investigation of thin layers of photosensitive material on a substrate, the most promising techniques are the capillary dilatometer and gas pycnometry [106], [107].

As a capillary dilatometer for a sample of several centimeters in size requires large amounts of mercury and is difficult to set up, gas pycntometry with nitrogen as analysis gas was chosen for the volumetric analysis. Gas pycnometers are based on Boyle's law, which states that $p \cdot V = const.$, when temperature and number of gas particles are constant. The schematic of a gas pycnometer is shown in Figure 4.6; the sample is placed inside a chamber with a pressure sensor and a known volume V_C. In the first step, the valve A is opened and the sample chamber, whose volume is reduced by the volume V_S of the sample, is filled to the initial pressure P_1, while the expansion chamber is at the atmospheric pressure P_A. Then, valve A is closed and valve B is opened, and the gas enters the expansion chamber of the known volume V_E. The machine then waits until the pressure is constant and records its value P_2. Under the assumption, that the temperature is constant and there is neither a gas leak nor gas evaporating from the sample, Boyle's law yields

$$(V_C - V_S)\cdot P_1 + V_E \cdot P_A = (V_C - V_S + V_E)\cdot P_2$$

$$\Rightarrow V_S = V_C - \frac{P_A - P_2}{P_2 - P_1}\cdot V_E.$$

As the experimental values are not absolute pressures, which are the subject of Boyle's law, but gauge pressures, the measured initial pressure is $P_I = P_1 - P_A$ and the final pressure is $P_F = P_2 - P_A$. With these definitions, the sample volume becomes

$$V_S = V_C - \frac{V_E}{P_I / P_F - 1} \tag{4.1}$$

Both the volume of the sample chamber and the volume of the expansion chamber have to be calibrated prior to any measurements. This is done using a calibration standard made from stainless steel with a known volume.

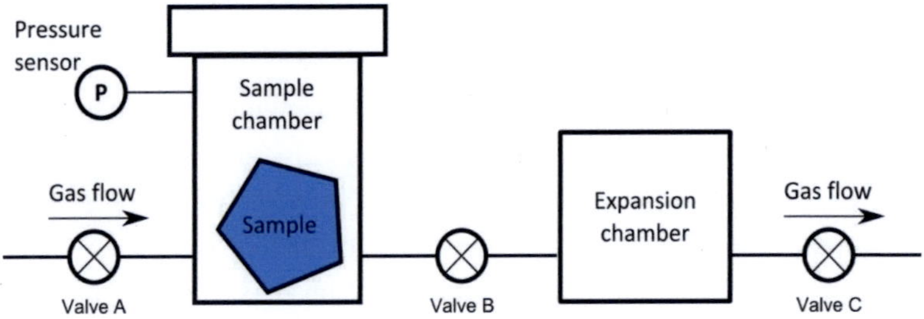

Figure 4.6: Schematic of a pycnometer

All measurements presented in this work were done using an *AccuPyc II 1340* device by the manufacturer micromeritics, with a custom modification of a sample chamber that can be heated up to 95 °C. Upon heating the chamber, the volume changes, which means that for each measurement temperature, a separate calibration must be carried out. The shape of the sample chamber is cylindrical, with a nominal volume of 100 cm^3. The inner diameter of the sample chamber is 46.2 mm, which restricts the sample size. A standard 4-inch wafer was cut into pieces with 32.5 mm maximum edge length to fit inside the chamber.

The absolute precision of the device is limited by the accuracy of the pressure sensor and the size of the sample chamber. The manufacturer guarantees a volume measurement precision of +/- 0.02 % of the sample chamber volume, and the device operates at its maximum precision when filled to about 2/3 of the sample chamber volume. A single piece of 32.5 x 32.5 mm^2 Si wafer with a thickness of 525 µm and a layer of 300 µm photoresist only has a volume of 0.87 cm^3, which means that the device will not operate at its optimum and the precision will only be about 2 % of the sample volume if a single sample piece is used. As the goal is to measure the shrinkage of the resist during PEB, which is expected to be in the range of 7 % of the resist volume, this precision is not sufficient for an accurate measurement. To overcome this problem, a sample holder was designed to fit several of the same 32.5 x 32.5 mm^2 wafer cutouts in the sample chamber for measurements. The dimensions of the sample holder, the wafer cutting scheme and an artist rendering of the loaded sample holder that was manufactured in aluminum by the IMT workshop can be seen in Figure 4.7. With this setup, 12 of the wafer cutouts were measured at the same time, the total resist volume was around 4 cm^3, resulting in a minimum precision of around 0.5 % of the resist volume of a single measurement, which should be sufficient to see the volumetric changes during the PEB. The precision can be increased by measuring the volume several times and averaging the results. In practice, 50 cycles were run for most measurements, and the standard deviation for these extended measurements varied between 0.004 cm^3 and 0.012 cm^3, with an average of 0.010 cm^3. The typical error of the value for the resist volume should thus be around 0.25 %.

The pycnometer measures the sum of the volume of all objects in the sample chamber; in order to get the volume of the resist only, a series of offset measurements has to be taken. The term 'measurement' refers to a volumetric measurement in the pycnometer with 50 – 300 cycles, the individual results are identified with the letter V, the subscript indicated and the temperature T of the measurement. The measurement procedure was conducted as follows:

1. Measurement of the empty sample holder
   - $V_{SH}(T)$
2. Measurement of substrates + sample holder at room temperature and PEB temperature
   - $V_{W}(T)$
3. Spin-coating and softbake of samples
4. Measurement of coated samples at room temperature
   - $V_{WR}(T)$
5. Exposure of samples
6. Measurement of exposed samples at room temperature
   - $V_{WRE}(T)$
7. Heating up pycnometer to PEB temperature

8. Measurement of volumetric change during PEB
   - $V_{PEB}(T)$
9. Measurement of final sample volume at room temperature
   - $V_{END}(T)$

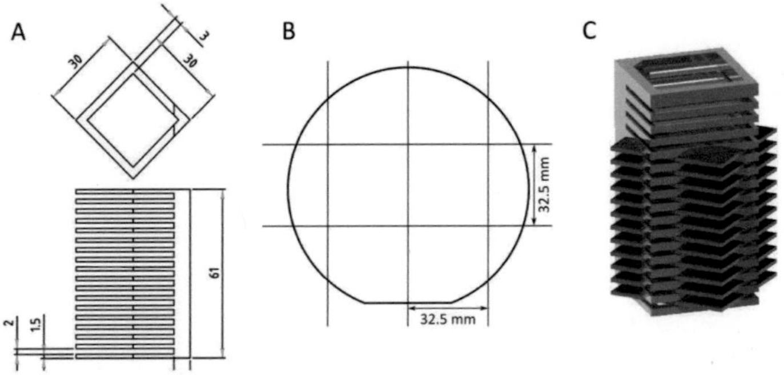

Figure 4.7: **A:** dimensions of the pycnometer sample holder; **B:** Wafer cutting scheme for Pycnometry samples; **C:** Artist rendering of sample holder with samples, from [108]

The resist volume V_R is obtained by $V_R(T)=V_{WR}(T)-V_W(T)$. This procedure theoretically allows the determination of the shrinkage of the resist by comparison of V_{WR}/V_{WRE} and V_{END}, as well as by directly looking at the volumetric change during the PEB. The exposures for the Pycnometry samples were done with UV light using a LH5 device by *Karl Süss* to be independent of the synchrotron time schedule. As pointed out in section 3.5.3, the polymerization mechanism is assumed to be the same in both cases.

The volume evolution during the PEB does not only provide a start and end value for the resist volume, but also allows quantification of the speed of the volume shrinkage. Figure 4.8 shows the evolution of the resist volume for a sample of mr-X 50 with a thickness of 300 µm during the standard PEB. As the heating of the pycnometer to the PEB temperature is very slow – heating up to 75 °C from room temperature can take several hours – the samples must be inserted into the heated sample chamber. The sample holder is also heated up, but obviously must be taken out of the chamber to be loaded. During the loading, the sample holder cools down; when inserting the holder with the samples into the chamber, both heat up again to the chamber temperature. During this heating process, the volume first increases, and then decreases as the resin starts to crosslink. This is visible in the experimental data as a rising volume at the beginning of the measurement. After this initial rise, an exponential of the form

$$V(t)=V_{END}+A\cdot\exp\left(-\frac{t}{\tau}\right)$$

is fitted to the data starting at the point of maximum volume to quantify the speed of the crosslinking reaction. The fit is executed via the parameters V_{END}, A and τ, the time constant τ provides the measure of the speed, the smaller it is, the faster the shrinkage is going. Volume shrinkage also is a measure of monomer conversion, the time constant of the volume shrinkage therefore also is a measure of the speed of the polymerization reaction [109].

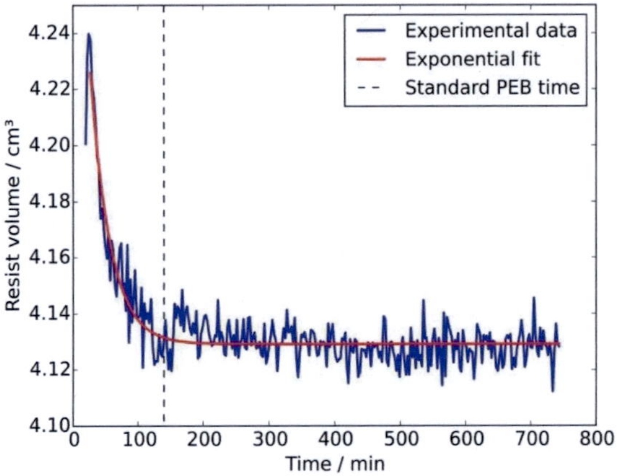

Figure 4.8: Evolution of the resist volume during the PEB at 75 °C. The dashed line represents the standard PEB time, the solid red line is an exponential fit to the experimental data

The dashed line in Figure 4.8 marks the standard PEB time; the major part of the shrinkage has already happened at this point, as it was also seen in the DSC measurements in section 4.2. However, the evolution of the resist volume suggests, that a higher degree of crosslinking would still be achievable by extending the PEB. The following sections show how the substrate storage time and the PEB temperature influence the course of the crosslinking reaction.

4.4.1 Variation of storage time

As the tensile testing experiments revealed a strong influence of the substrate age on the mechanical properties of the final structures, an analogue set of samples was also prepared for the measurement of the shrinkage with pycnometry. A series of 4 sample sets, each consisting of 12 wafer cutouts of 32.5 x 32.5 mm^2 was prepared with a layer of ca. 300 µm resist from the same bottle (mr-X 50, K102). The samples were stored for 1, 5, 20 and 33 days before exposure with UV light. The PEB was done inside the pycnometer for all samples at a temperature of 75 °C. For all of the samples, a fit like presented in Figure 4.8 was performed. The results for the time constant τ are plotted in Figure 4.9; the errorbar signifies the uncertainty of the fit parameter. A clear trend to shorter time constants is visible with increasing storage time, τ falls from 37 min for the "fresh" sample with a storage time of only one day to 13 min for the sample with a storage time of 33 days. A time constant of 13 min means that after $4 \cdot 13\ \text{min} = 52\ \text{min}$, more than 98 % of the shrinkage has happened, whereas the same percentage of shrinkage would take 148 min for the fresh sample. This is longer than the standard PEB time of 120 min, which in this case would still result in 96 % of the possible shrinkage at this temperature has happened. These results provide an explanation for the supposed lower degree of crosslinking in fresh samples derived from the mechanical testing – a larger time constant means that in the fixed PEB time, a lower degree of crosslinking is achieved.

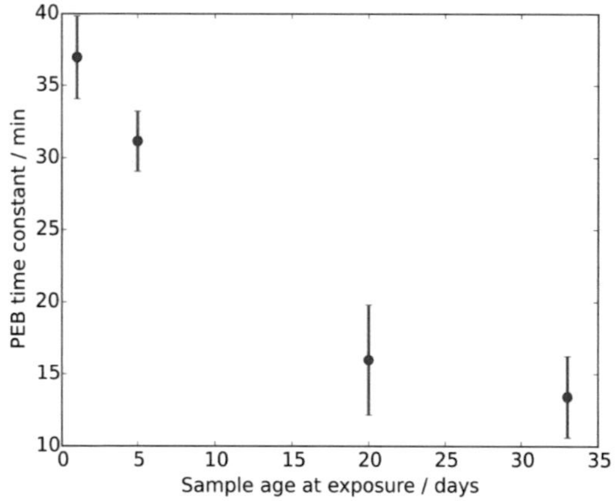

Figure 4.9: Time constant τ of fit to shrinking resist volume during PEB at 75 °C for samples with different storage time between spin-coating and exposure

The most important results apart from the PEB time constant are the absolute shrinkage of the resist as measured during the PEB, $\Delta V(75\ °C) = 1 - V_{R,END}(75\ °C) / V_{R,MAX}(75\ °C)$. The values are compiled in Table 4, a shrinkage between 1.5 % and 3.3 % has to be compared to the reported value of 7.5 % for SU-8 [90]. A lower value than for SU-8 is expected, as in previous studies, the tension introduced by the crosslinking of SU-8 and mr-X was compared by measuring the deformation of a glass substrate with resist coating, the values for SU-8 ranged from 11 MPa to 16 MPa, while for mr-X, a tension of 6 +/- 3 MPa was found [81], [88]. In the pycnometry data, there seems to be a trend towards lower shrinkage with increasing substrate storage time; the values for the three samples with the shortest storage time agree within their errors, while the samples with 20 and 33 days storage time show less shrinkage. The error for ΔV(75 °C) is realtively high, because the volume $V_{R,MAX}$(75 °C) is determined with a single measurement, while the other values rely on averaging multiple measurements.

Table 4: Resist volumes measured at different process steps with varying storage time

Sample age / days	V_R (23 °C) / cm³	$V_{R,END}$ (23 °C) / cm³	ΔV (23 °C) / %	$V_{R,MAX}$ (75 °C) / cm³	$V_{R,END}$ (75 °C) / cm³	ΔV (75 °C) / %
1	3.523 +/- 0.010	3.558 +/- 0.010	-1.0 +/- 0.4	3.710 +/- 0.023	3.609 +/- 0.014	2.7 +/- 0.7
5	4.111 +/- 0.008	4.103 +/- 0.010	0.2 +/- 0.3	4.322 +/- 0.020	4.180 +/- 0.009	3.3 +/- 0.5
20	3.309 +/- 0.010	3.359 +/- 0.008	-1.5 +/- 0.4	3.443 +/- 0.026	3.389 +/- 0.021	1.6 +/- 1.0
33	3.563 +/- 0.010	3.318 +/- 0.015	6.9 +/- 0.5	3.683 +/- 0.021	3.628 +/- 0.010	1.5 +/- 0.6

Another way to quantify the shrinkage would be to compare the resist volume at room temperature before and after PEB, $\Delta V(23\,°C) = 1 - V_{R,END}(23\,°C) / V_R(23\,°C)$. However, there is no consistent outcome of this procedure, some samples even show an increase in volume after PEB. These results are likely incorrect, because volume shrinkage during PEB was observed for all samples. The errors of the individual measurements are as expected and do not account for the large discrepancy to the expected values, which means that there must be a systematic error. The most plausible reason is that before PEB, the samples tend to outgas, especially solvent molecules. Consequently, the assumption of constant number of particles is not valid any more, and the final pressure P_F of the measurement will be higher than without the emission of extra particles. According to equation (4.1) on page 60, this leads to a result for the sample volume which is smaller than the real volume. After PEB, this outgasing is significantly reduced, and therefore these two measurements cannot be compared due to different systematic errors.

Variation of PEB temperature

A major contribution to stress in crosslinked resist structures comes from different thermal expansion coefficients of resist and substrate and is thus thermally induced [110]. A reduction of the PEB temperature is desirable to reduce the resulting stress in the structures which can lead to deformations and breaking of structures in the worst case. A change of the PEB temperature is expected to have an impact on the speed of the crosslinking reaction and might also affect the shrinkage, because a lower temperature may lead to less crosslinking. To test this influence, a set of 5 samples was prepared from the same resist bottle (mr-X 50; K116) using the same spin-coating and soft bake parameters as for the storage time series. All samples were stored for six weeks between spin-coating and exposure and were exposed with the same UV dose. The PEB temperature was varied according to the results from the DSC measurements from 60 °C to 80 °C, see Figure 4.1 B.

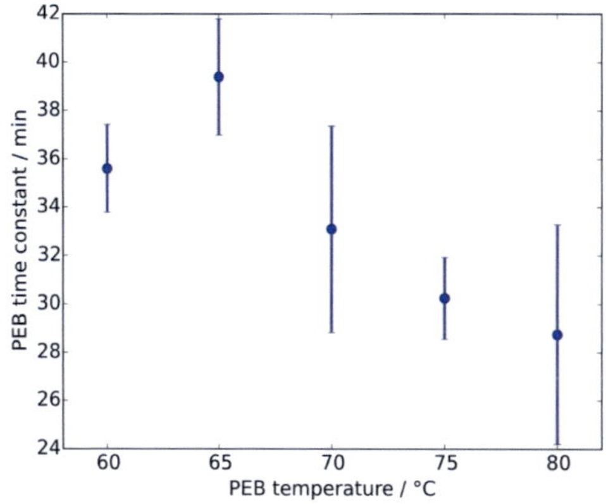

Figure 4.10: Time constant τ of exponential fit to shrinking resist volume during PEB at varying temperature

The results for the PEB time constants are plotted in Figure 4.10. The expected trend to a faster reaction with increasing PEB temperature is apparent, only the sample with a PEB temperature of 65 °C shows a larger time constant than the sample with 60 °C instead of a smaller one. An explanation could be that the PEB temperature of 60 °C only induces partial crosslinking and seems faster because it reaches a steady state at a lower degree of crosslinking. To test this hypothesis, both samples were developed for 1 h in PGMEA and rinsed in Isopropanol. Their weight was tracked to see any loss of material, but no change in weight was found in either of the samples, indicating that both were crosslinked enough to withstand the development. This means that the difference in degree of crosslinking cannot be very large, and likely one of the measurement points is an outlier. The sample with a PEB temperature of 75 °C shows a much larger time constant as the sample with 33 days storage time from the previous sample set, which is unexpected. The only known difference between the two samples is that the resist comes from two different fabrication batches, which could explain varying behavior. This indicates that it is difficult to compare the results from different batches. One has to be aware that the differences from batch to batch could be higher than the differences due to changed process parameters.

Table 5: Resist volumes measured at different process steps with varying PEB temperature

PEB temp / °C	V_R (23 °C)/ cm³	$V_{R,END}$ (23 °C) / cm³	ΔV (23 °C) / %	$V_{R,MAX}$ (75 °C) / cm³	$V_{R,END}$ (75 °C) / cm³	ΔV (75 °C) / %
60	3.913 +/- 0.008	3.925 +/- 0.011	-0.3 +/- 0.3	4.052 +/- 0.026	3.941 +/- 0.017	2.7 +/- 0.8
65	3.940 +/- 0.012	3.979 +/- 0.008	-1.0 +/- 0.4	4.130 +/- 0.024	4.021 +/- 0.016	2.6 +/- 0.7
70	4.454 +/- 0.009	4.465 +/- 0.016	-0.2 +/- 0.4	4.218 +/- 0.028	4.142 +/- 0.020	1.8 +/- 0.8
75	4.025 +/- 0.007	4.078 +/- 0.014	-1.3 +/- 0.4	4.239 +/- 0.022	4.127 +/- 0.010	2.6 +/- 0.5
80	3.997 +/- 0.009	4.062 +/- 0.011	-1.6 +/- 0.4	4.190 +/- 0.023	4.121 +/- 0.016	1.6 +/- 0.7

Table 5 summarizes the resist volumes at the different process steps for the samples with varying PEB temperature. For this set, all the volumes measured at room temperature after the crosslinking are larger than the volumes measured before exposure. The shrinkage ΔV(75 °C) is comparable to the values measured at constant PEB temperature before, but no trend is visible that is larger than the measurement error.

4.4.2 Modified PEB for lithographic process

With the result that crosslinking the structures at 60 °C instead of the standard 75 °C yields a comparable shrinkage at just a moderate increase of reaction time, the effect of this parameter change on X-ray lithographic structures needs to be studied. A low PEB temperature is desirable, because it reduces stress induced by the mismatch in the coefficient of thermal expansion (CTE) between the resist and the substrate. While the CTE for the standard substrate silicon is $2.6 \cdot 10^{-6}\ K^{-1}$, the coefficient of crosslinked SU-8 was measured to be $52 \cdot 10^{-6}\ K^{-1}$ [111]. Uncrosslinked epoxy resins have a significantly higher CTE, the difference can be a factor of 5 [112]. To judge whether a lower PEB temperature increases quality of grating structures, a layout was chosen that gave unsatisfactory results with the standard parameters. A mask with a period of 2.19 µm and a duty cycle of 0.66 was used, leading to a width of the transmitting mask lines of about 750 nm. The resist height was chosen as 10 µm (AR 13.3). To ensure that no influence from the preprocessing disturbs the results, one sample was prepared and then exposed with 140 J/cm³ bottom dose at LIGA I. After exposure, the wafer was broken in parts that were subjected to PEB with different parameters. One part was subjected to the standard PEB at 75 °C in a vacuum oven and one to a PEB at 60 °C on a hotplate. To track possible differences between oven and hotplate, another part was subjected to PEB on a hotplate at 75 °C. To account for the slower reaction time, the duration of the PEB at 60 °C was increased to 4 hours, additionally the time for heating up from room temperature to the PEB temperature was increased to 2 hours. The 75 °C PEB on the hotplate was done with the same timing as in the oven, i.e. heating in 20 min, hold PEB temperature for 2 hours.

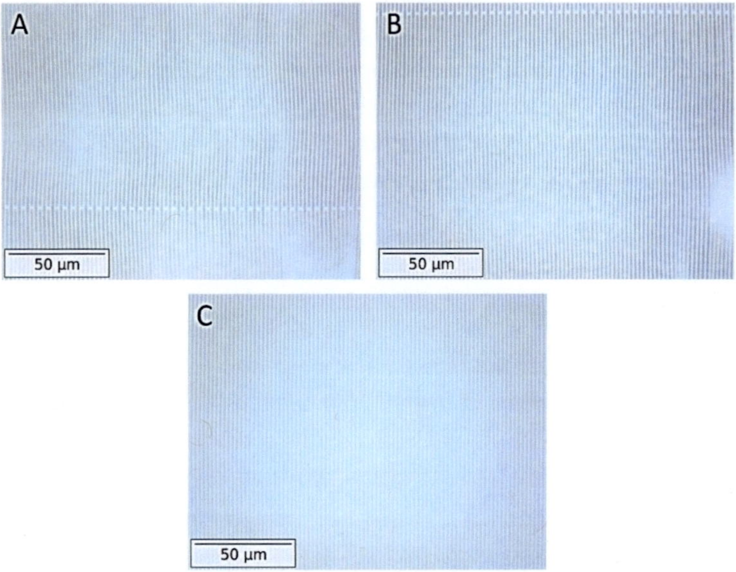

Figure 4.11: Comparison of 2.19 µm period grating structures after PEB with varied parameters: 75 °C in vacuum oven (**A**), 75 °C on hot plate (**B**), 60 °C on hot plate (**C**)

The outcome of the three samples can be seen in the representative microscope images taken after PEB of the sample with 140 J/cm³ bottom dose, shown in Figure 4.11. While the lines in both Figure 4.11 A and B, which represents the PEB at 75 °C in the vacuum oven (A) and on the hot plate (B), appear wavy and are of comparably poor quality, the lines in the image C are straight, the pillow-like deformation at the edges only stems from optical aberrations in the imaging system. A repetition of this procedure with a sample with a higher bottom dose (180 J/cm³) yielded better structure quality in the parts with a 75 °C PEB, but also showed the best results in the part with 60 °C PEB. Furthermore, the structure quality was preserved after development, but freeze drying is necessary in all cases to prevent lamella collapse. Thus, it can be concluded that the shift to a lower PEB temperature and longer PEB duration is beneficial for structure quality in small periods. It is expected that this improvement also translates to larger structure sizes that show waviness after PEB. It also suggests that the waviness is not introduced by the stress introduced by the chemical shrinkage of the resist, since the amount of shrinkage measured was similar for all PEB temperatures, but instead is governed by thermal stress, which is reduced when using lower PEB temperatures. Note that this is specific to the mr-X resist and cannot be directly transferred to standard SU-8, because of its higher chemical shrinkage. Nevertheless, minimizing the PEB temperature should also lead to less deformations in the case of standard SU-8.

4.5 Summary of photoresist characterization

The study of the thermomechanical properties of the photoresist under different processing conditions showed a large influence of the storage time between coating and exposure of the sample on the Young's modulus and tensile strength, with the older samples appearing stiffer than fresh samples. This was underlined by investigation of the crosslinking reaction speed measured by gas pycnometry, where the older samples showed a shorter time constant of volumetric shrinkage which results in a higher degree of crosslinking after the fixed PEB time. It was found, however, that a second softbake step at lower temperature (40 °C) can have the same effect on the mechanical properties as the storage time and can thus be used to shorten the waiting time for a sample. The samples with a second softbake also showed a more brittle behavior than the standard samples, which hints to a lower overall solvent content. The solvent content and distribution therefore appears as the key parameter in the change of mechanical properties during the substrate ageing. The tensile testing also confirmed that with increasing exposure dose, the Young's modulus, tensile strength and elongation at break of the samples are increased in a similar way as by a hard bake.

The volumetric measurements also revealed that the crosslinking reaction speed depends on the PEB temperature, but no significant change in the absolute shrinkage was found after long waiting times, indicating that the PEB temperature can be lowered when the duration is increased. Doing so resulted in grating structures with strongly reduced waviness and thus increased quality, which was shown as an example with grating lines of 800 nm width, 10 µm height and 500 µm length. These results also show that the wave-like deformations are not caused by the chemical shrinkage, but by thermally generated stress. Reducing the PEB temperature to 60 °C while increasing the duration to at least 4 hours should therefore be integrated as standard to the process, especially for structural dimensions that are prone to waviness, like long lines with an aspect ratio above 10.

5 Grating interferometry with low absorption substrates

In the previous chapter, it was shown that grating structure quality can be enhanced by adjusting the process parameters. The structure quality is the main determining factor for the achievable visibility in a grating interferometer. However, the sensitivity of a grating interferometer is not only determined by the visibility of the interference pattern, but also by the number of detected photons (see section 2.2.3). The substrates of the gratings play an important role for this aspect, because they act as absorbing filters for the radiation. The effect of additional, unwanted absorption in the substrate depends on the grating position. If the grating is upstream of the sample, the absorption decreases the available flux and thus causes longer exposure times. If the grating is downstream of the sample, the photons absorbed in the substrate carry information from the sample that is lost; the signal-to-noise ratio is decreased and the necessary dose applied to the sample is increased, which can be problematic for radiation-sensitive samples in biomedical imaging.

The standard substrate in the direct LIGA process is a 4-inch silicon wafer with a thickness of 525 µm. At 30 keV, this substrate has a transmission of 85 %, which is still acceptable, but with decreasing energy this gets worse, at 10 keV only 2 % transmission are left and interferometry becomes practically impossible. An easy way around this for moderate energies is the use of thinner silicon wafers, and wafers with a thickness of 200 µm are now routinely used along with the 525 µm thickness, although their fragility decreases the fabrication yield. Silicon can also be thinned by a variety of techniques after the fabrication of the grating structures, but such an approach complicates the process and also reduces the yield [113]. A more promising way is to replace the silicon substrate with a less absorbing material.

In the low energy regime (<30 keV), where the substrate transmission becomes important, the mass attenuation coefficient μ/ρ of a material is approximately proportional to the third power of the atomic number Z

$$\mu/\rho \propto Z^3.$$

Any material with a lower atomic number than silicon (Z = 14) is therefore a candidate for an alternative grating substrate. The main criteria for grating substrates from a fabrication point of view are pointed out in section 3.1; Table 6 shows how different light materials fulfill them. Along with the light elements Beryllium and Boron, Graphite and several polymers were investigated. Graphite is the most interesting form of Carbon for the purpose of grating substrates because of its intrinsic conductivity; as polymers we considered Polyimide, Polyetheretherketone (PEEK) and Polycarbonate. Polyimide is very resistant both to chemicals and ionizing radiation and is therefore used in many radiation related applications, e.g. as filter or as vacuum window, PEEK is a semi crystalline thermoplastic polymer that has a high resistance against most chemicals and is often used in medical applications due to its high biocompatibility and Polycarbonate is a lighter and much cheaper alternative to the two which is also comparatively resistant to the chemicals used in the process.

The best rating of the materials in Table 6 is achieved by beryllium (Z =4), followed by boron (Z = 5), but as beryllium is highly toxic and boron is rare and therefore expensive, both are excluded from consideration. Figure 5.1 shows the maximum affordable substrate thickness for the other materials when imposing the condition that at least 90 % of incoming photons should be transmitted. It is obvious that from the point of view of transmission, it is highly favorable to use graphite and the polymers instead of silicon, the thickness difference is roughly one order of magnitude. All four materials were tested for fabrication, but only graphite and polyimide were available with sufficient quality for processing, especially in terms of substrate planarity. A substrate with insufficient planarity will lead to an inhomogeneous resist thickness after soft bake and is thus an exclusion criterion.

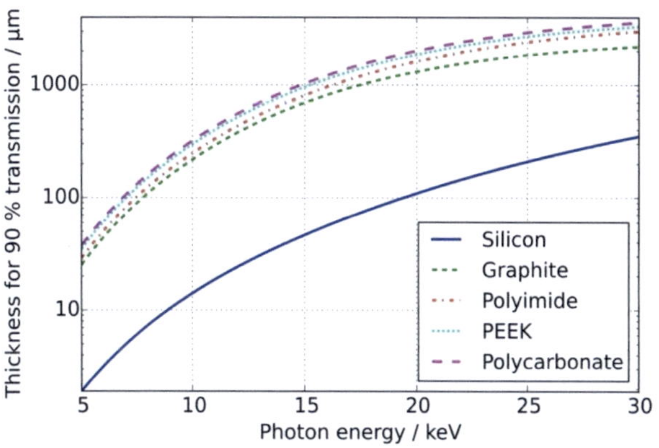

Figure 5.1: Maximum substrate thickness to retain 90 % transmission per grating for different materials on a semi logarithmic scale, based on data from [114]

5.1 Polyimide

Polyimide was used in three different thicknesses: a 125 µm foil (DuPont Kapton®) and plates of 500 µm and 1000 µm thickness (DuPont Vespel SEK®). For grating fabrication on polyimide substrates, they first have to be coated with an electroplating seed layer. This was done using an evaporation process to deposit a layer of 7 nm chromium and 50 nm gold either at IMT or at the commercial supplier Micromotive GmbH, Mainz, Germany. This seed layer is optimal for good height control, but the large X-ray absorption coefficient of gold leads to more secondary electrons from the substrate during exposure, which is especially limiting for the exposure of high structures at the LIGA II beamline. However, this is not of importance here, since gratings on low absorption substrates are important only for low imaging energies, where a gold thickness of up to 100 µm, which can be fabricated at LIGA I, is already sufficient.

Table 6: Evaluation of different materials for the use as grating substrates

	Mechanical stability	Chemical resistance	Radiation hardness	Thermal stability	Conductivity	X-ray transmission	Other characteristics
Silicon	+	++	++	++	-	-	Excellent surface quality
Beryllium	++	++	++	++	++	++	Toxic
Boron	++	++	++	++	+	++	High price
Graphite	+	++	++	++	0	++	Porous
PEEK	+	+	0	++	-	++	Insufficient planarity
Polyimide	++	++	+	++	-	++	
Polycarbonate	+	+	0	+	-	++	Insufficient planarity

While the Vespel plates are sufficiently stable for processing, the foils have to be bonded to a carrier during the processing. This is possible using e.g. Crystalbond 509 [115], dissolved in 80 %wt Acetone and spin-coated onto a silicon wafer with 2000 rpm for 20 s and then soft baked at 90 °C inside a vacuum oven. To ensure bubble-free bonding, it is preferable to do the actual bonding process in a vacuum environment, which was done with a wafer bonding machine at the Institute of Applied Materials, Institute for Material Process Technology at KIT Campus north. After processing, the foil can be released from the substrate by either dissolving it in acetone or thermally by heating to temperatures over 70 °C. To minimize thermal stress on the structures, all samples were debonded using acetone.

Apart from the preparation processes just described and the debonding for the foils after electroplating, there is no need for change in the process compared to substrates with a CrAu layer on top of a silicon wafer. It was seen, however, that when attempting to remove the resist after electroplating with oxygen plasma, the CrAu layer can be sputtered off the substrate, exposing the PI foil to plasma induced damage. This can lead to deformation of the foil, which is less severe for the thicker plates, but still present. Thus, partial resist stripping must be performed with caution and full stripping is not advised for PI substrates.

5.2 Graphite

As graphite is already conductive, no electroplating seed layer is necessary. Graphite wafers were purchased from Ohio Carbon Blank Inc. [116] with a thickness range from 200 µm to 1000 µm. The material is synthetic graphite with a porous structure and an average grain size of 1 µm (Supplier identifier EDM-AF5). Due to the porous structure, it soaks up the photoresist during spin-coating, which makes height control difficult, but at the same time provides excellent adhesion of the resist to the substrate and effectively prevents leaking of electrolyte under the lamellae in electroplating.

The conductivity of graphite along with its low X-ray absorption offer another advantage, which is that far less secondary electrons from the substrate are to be expected during the X-ray exposure step when compared to standard substrates with either a titanium or gold electroplating seed layer. This means that samples on graphite wafers are less prone to unwanted resist crosslinking near the substrate surface, which is a problem for metal layers [117].

Due to their microstructure, the graphite wafers produce a lot of small angle scattering of the incident beam. Small angle scattering is a problem in an interferometer, because it reduces the coherence of the illumination and introduces noise. However, for the practical implementation, it is important at which point of the interferometer the scattering happens. The sensitivity of a Talbot-Lau interferometer is highest at the position of the phase grating G1, and scales linearly with the distance of the sample to G1 until it reaches zero at the positions of G0 and G2 [118]. Studies have shown that consequently, at the positions of G0 and G2, a strongly scattering substrate does not have a significant influence on the visibility of the interferometer, while at the position of G1, it might reduce the visibility to less than 50 % of its original value [94]. With these results, it is obvious that graphite cannot be used as phase grating substrate, but is a candidate for the absorption gratings G0 and G2.

5.3 Test samples and imaging results

A series of phase and absorption gratings with PI and graphite substrates was produced and showed good structure quality under microscopic inspection. However, this is not sufficient to assess the actual grating performance, as it only provides information about the surface. Therefore, two different imaging setups were used to show different aspects of the effects of gratings on low absorbing substrates, a three-grating setup at the Technical University of Munich (TUM) using a rotating anode tube source and a two-grating setup built at the beamline IMX at the Brazilian Synchrotron radiation laboratory (LNLS).

5.3.1 Sensitivity

The increase in transmission using low absorbing substrates is expected to increase the sensitivity of an interferometer. To test this influence, a symmetrical three gratings setup with a grating period of $p_0 = p_1 = p_2 = 5.4$ µm at TUM was used. The source was an Enraf Nonius FR 591 rotating anode with a tube voltage of 40 kV, the design energy of the setup was 27 keV and a Dectris Pilatus II 100k photon counting detector was used; a more detailed description of the setup can be found in [119]. The setup was equipped with gratings on silicon substrates (2x 200 µm, 1x 525 µm) as a reference, which were subsequently replaced by a set of gratings on PI substrates (1000 µm for the absorption gratings, 125 µm for the phase grating) and a set of gratings on 200 µm graphite substrates, which were combined with the phase grating on PI foil. As the absorber heights for all gratings were around 70 µm, a similar visibility is expected and was found for all grating sets, the mean visibility with both the silicon and the graphite set was 24 %, with the PI set it was reduced to 21 %. This small difference can be explained by a slightly inferior structure quality in one the absorption grating used as G0, which showed periodic ruptures.

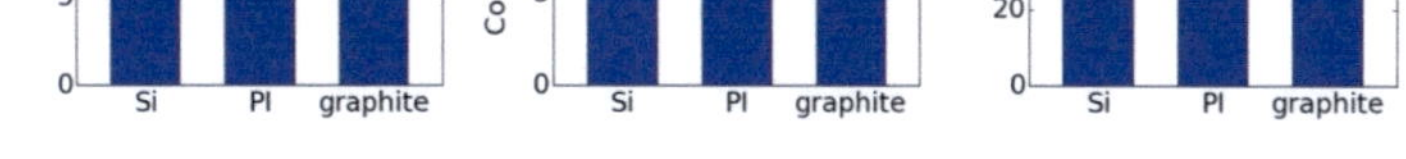

Figure 5.2: Visibility, counts per second and sensitivity expressed as minimum resolvable refraction angle for different grating sets, from [120]

Figure 5.2 shows the values for visibility, counts per second and minimum resolvable refraction angle as calculated from $\alpha_{min} = \frac{p_2}{2\pi d} \cdot \frac{2}{V\sqrt{N}}$ (see equation (2.13) in section 2.2.3), with the inter-grating distance $d = 857$ mm and the respective values for the visibility *V* and photon counts *N,* with an assumed exposure time of 1 s per step. The sensitivity for the silicon grating set was 112 nrad, 72 nrad with the PI set and 58 nrad with the graphite set, which means that the sensitivity could be nearly doubled by changing the substrate material while keeping all other setup parameters constant [120].

5.3.2 Low energy imaging

With the low absorbing substrates, grating interferometry at energies below 10 keV becomes feasible, which could be shown by experiments performed at the beamline IMX of the Brazilian Synchrotron Light Laboratory (LNLS) in Campinas, Brazil. A multilayer monochromator was used to select an energy of 8.3 keV from a bending magnet source. The source size is approximately 391 µm x 97 µm (horizontal x vertical), and the phase grating was placed 17 m downstream of the source. The grating lines were oriented horizontally to make use of the smaller vertical source size and enable the use of a two-grating interferometer. The phase grating G1 had a period of 2.4 µm, with $\pi/2$ phase shifting lines that were made from mr-X with a thickness of 10 µm to ensure high transmission also in the phase shifting lines, the substrate was 1000 µm PI. The absorption grating G2 had the same period, with absorbing lines of ca. 30 µm Au, which leads to an absorption of over 99 % of photons. As substrate, the 125 µm PI foil was used. Together, the substrates lead to a transmission of 43 % at 8.3 keV. The inter-grating distance was set to 97 mm, corresponding to the 5th fractional Talbot distance; the sample was placed 135 mm upstream of G1. Detection was done using a pco.2000 camera mounted behind a microscope objective and scintillator, yielding an effective pixel size of 1.64 µm and a field of view of 3.4 x 3.4 mm^2.

The whole experimental setup is shown in the photograph in Figure 5.3. The phase grating was mounted on a linear piezo and had to be inclined to match the magnified period of the phase grating to the unmagnified period of the analyzer. Unfortunately, this axis could not be motorized and the alignment had to be done by hand with a wedge that was placed under the piezo mounting.

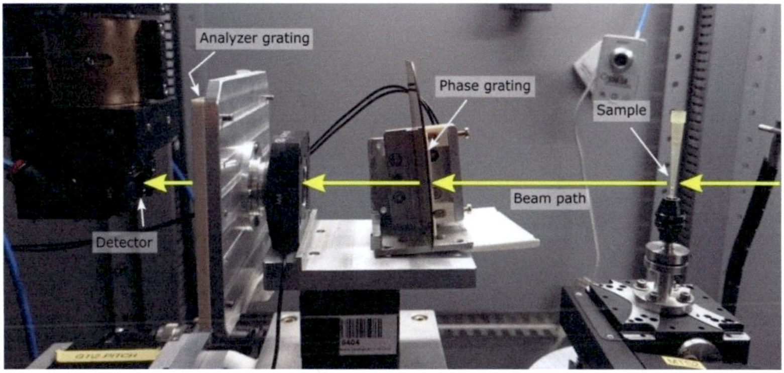

Figure 5.3: Experimental Setup at the IMX beamline of the LNLS with two gratings on polyimide substrates

Using an exposure time of 30 s per step over 8 steps, the mean visibility over the whole field of view was 40%, mostly limited by setup instability and grating imperfections (Figure 5.4); the inhomogeneous distribution of the visibility is mostly due to the profile of the beam coming from the multilayer monochromator. The sensitivity as determined from the standard deviation of the differential phase signal in a region of interest without sample was approximately 165 nrad.

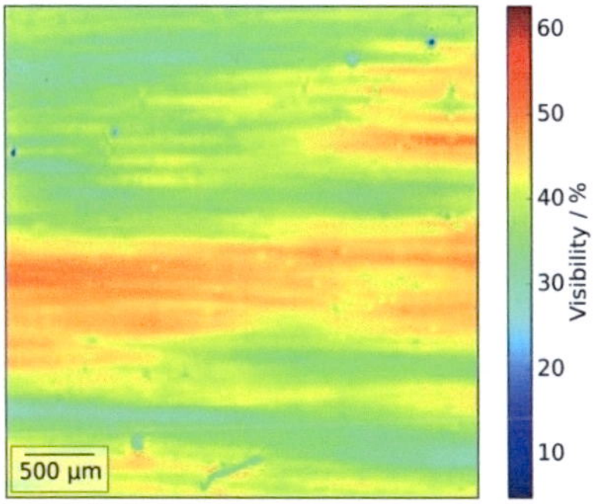

Figure 5.4: Visibility map obtained with the two-grating interferometer at 8.3 keV at the IMX beamline of the LNLS

When taking an image, the differences in intensity are normalized with the flat field correction and the inhomogeneous visibility map only results in a variation of the sensitivity. Figure 5.5 shows an example of a transmission, differential phase and dark field image taken with the interferometer equipped with gratings on low absorbing substrates. The sample is a capillary with polymer microspheres that are filled with magnetic γ-Fe_2O_3-nanoparticles, synthesized by the group of Prof. Sidney Ribeiro at State University of Sao Paulo, Brazil. The particles show both a strong absorption and a significant dark field signal, the profile of the visibility map seen in Figure 5.4 is partly seen also in the dark field signal, but hardly visible in the other modalities. The results show that the gratings on low absorbing substrates enabled low energy grating interferometry at the IMX beamline of the LNLS.

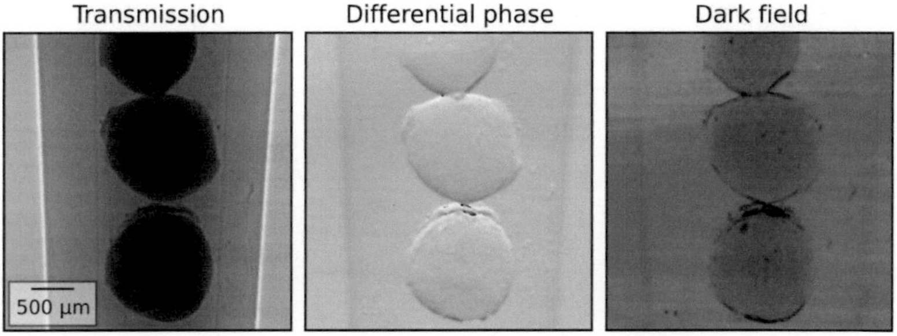

Figure 5.5: Transmission, differential phase and dark field image of polymer microspheres with Fe_2O_3-nanoparticles, taken with the two-grating interferometer at 8.3 keV

6 Quantitative analysis of X-ray lenses

Just as it is not sufficient to grade the structure quality of a grating using optical and electron microscopy, these methods can only give a rough idea about the performance of a compound refractive lens (CRL). Characterization of size and shape of the focal spot with a knife edge test is an easy way to assess a lenses quality in the beam. However, this type of test provides very little information on the type of aberration and their causes. Absorption based radiography is limited in the case of lenses because of their high transmission, which makes it difficult to achieve a reasonable signal-to-noise ratio. Time consuming techniques such as micro- or even nano-computed tomography or laminography can yield a 3d image of a lens and reveal voids and inclusions in the material, but this type of measurement is not suited for high throughput quality control [121]. Other types of setups are suitable for lens analysis, such as a Ronchi type interferometer or the analysis of the image of a grating in a microscopy setup, but these setups do not easily provide quantitative data [122], [123]. Another interesting alternative is the use of ptychography, which can provide an image of the wavefront behind a complete CRL, but is time-consuming and numerically expensive [124].

Grating based phase contrast imaging, as described in section 2.2, provides an easy and efficient way to measure the wavefront phase gradient, and is therefore well suited for the characterization of optics [51], [52]. In this work, a dedicated interferometer for the quantitative characterization of refractive X-ray lenses was set up at the beamline ID06 of the European Synchrotron Radiation Facility (ESRF) and employed for the analysis of CRLs fabricated by X-ray lithography and their comparison to other lenses. The following sections describe the design and setup of the interferometer, as well as the methods used for data analysis, which are explained using the example of a lens fabricated by beryllium imprinting [125]. This is followed by an analysis of the defects and aberrations found in lithographic lenses and their modeling and explanation using the results from chapter 0.

6.1 Experimental setup

The design of the interferometer must consider the properties of the source, which in the case of ID06 is an undulator with a source size of 415(h) x 8.6(v) μm^2 rms. The space for building the setup in the experimental hutch is ca. 55 m downstream of the source. The design energy of the interferometer was chosen as 17 keV, because it is close to the K_{α}-emission of Molybdenum, a common cathode material for X-ray tubes that could be used for future transfer of the methodology to a laboratory source. The high monochromatic flux available from the undulator makes it possible to use a camera with small pixel size and still have short exposure time; consequently, a pco.2000 camera mounted behind a scintillator screen and microscope optics with an effective pixel size of 0.74 µm was used for detection. Using this high-resolution camera enabled the setup of a single grating interferometer, which greatly simplifies the alignment.

The grating was chosen as a $\pi/2$-shifting phase grating with a period of 10 µm, creating an interference pattern of the same period, which can easily be resolved by the camera. A piezo drive was employed to perform phase stepping to make use of the full detector resolution. The setup can also be operated in single shot mode, but the analysis methods for these images, such as the evaluation of spatial harmonics [126] or the cross-correlation of reference and sample image [127] sacrifice spatial resolution in favor of short acquisition times and low dose which are not of major importance for optics analysis.

The distance between grating and detector was chosen experimentally by moving the grating in small steps along the beam direction and analyzing the contrast of the recorded interference pattern. The distances found to yield the highest contrast were 0.2 and 0.7 times the Talbot distance $z_T = 2p^2/\lambda = 2.74$ m, with respective visibility of 50 % and 46 %. The theory predicts the positions with the highest contrast to be the fractional Talbot distances 0.25 z_T and 0.75 z_T. Different effects could be responsible for this discrepancy; the finite absorption and error in height of the grating lines, the duty cycle, which was measured to be 0.42 in the grating used here instead of the usually assumed 0.5, resulting in a different phase shift than expected and the finite divergence of the beam. To find out whether one of these factors could explain the findings, numerical simulations were carried out using a script based on Fourier-Transform Fresnel propagation. The Talbot carpet was simulated with a lateral precision of 1024 pixels per period and then binned to 14 pixels per period, corresponding to the situation at ID06. The maximum detector counts were set to 1000, which corresponds to the experimental values, additionally a dark count of 100 per pixel was assumed. The finite source size was implemented as a convolution with a Gaussian; the divergence was not considered because it would only shift the Talbot distance to larger values instead of the shorter ones experienced in the setup [46]. For the resulting Talbot carpet, the visibility was calculated along the propagation direction. It is plotted in Figure 6.1, with variation of the parameters duty cycle, absorption and phase shift.

The most asymmetric effect comes from the duty cycle, and a duty cycle of 0.4 shifts the maxima very close to the experimentally observed ones, which are marked with vertical lines in the plots. Even with a 0.5 duty cycle, the maximum is not at 0.25 and 0.75 times the Talbot distance z_T; this effect is entirely due to the dark counts and disappears when not accounting for them. A larger absorption has the same asymmetry, but a smaller effect than the duty cycle; a larger phase shift mainly lowers the contrast at the fractional Talbot orders, but hardly introduces asymmetry. Both a larger phase shift and a higher absorption are to be expected, as the grating lines were a little higher than intended. The discrepancy in the distances with the highest contrast can therefore be attributed to a combination of the three factors with the duty cycle as the leading contribution.

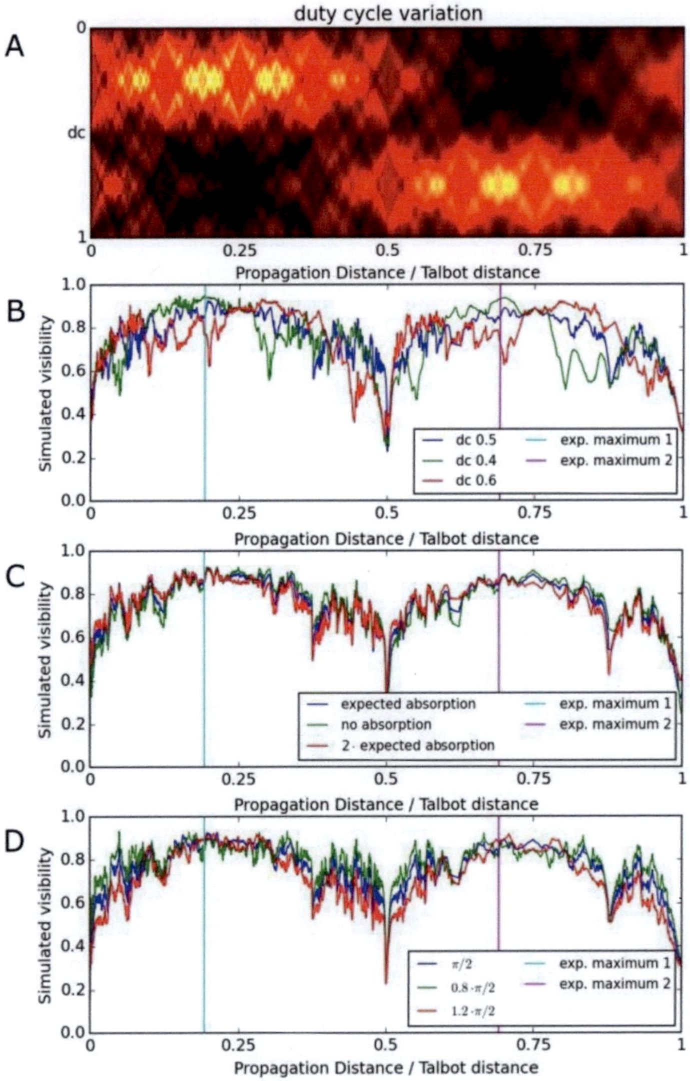

Figure 6.1: Simulated Talbot Carpet (**A**) and visibility along the propagation direction with variation in different phase grating properties: duty cycle (**B**), absorption (**C**) and phase shift (D)

For all experiments reported here, a one-dimensional grating was used, with the lines oriented horizontally. This orientation was chosen to make use of the much smaller source size, i.e. higher spatial coherence, in the vertical direction. A setup like this is only sensitive to the phase gradient in the vertical direction; for a full characterization of the wavefront, either the sample or the grating has to be rotated around the optical axis. Rotating the grating would be equivalent to the use of a 2d-grating and would lead to highly anisotropic sensitivity in the case of ID06; consequently, the sample was mounted on a rotational stage whose axis was aligned to the optical axis. A set of two phase stepping scans with a relative angle of $\Delta\theta = 90\,°$ is enough to fully characterize the wavefront. However, to increase robustness and to identify possible artefacts, scans at multiple angles were performed, in sets of two with a relative angle of 90 °. One set was then arbitrarily chosen as the reference directions x ($\theta_x = 0\,°$) and y ($\theta_y = 90\,°$), and the other sets were projected on these axes and normalized by the number of sets to form the final image. To achieve correct superposition of the image sets, they have to be aligned after acquisition, which was done by rotating the transmission images and using cross-correlation to detect the lateral shift.

6.1.1 Focal length limits for quantitative analysis

The interferometer design described here is limited in the focal length of a lens to be analyzed for different reasons. Obviously, the focal length f must be longer than the sample to detector distance d_s in order to still get an image of the aperture. It would be feasible to place a lens with a shorter focal length to a distance where the detector is behind the focus, but in such a configuration, there is no possibility to obtain a reference image of the grating, and therefore a quantitative analysis would no longer be easily available.

To be able to analyze the interference pattern, restricting the focal length to $f > d_s$ is not sufficient, because the demagnified intensity modulation still must be resolved by the camera. For the following calculations, we set $d_s = d + a$, with d being the distance from grating to detector and a the distance of sample to grating, which was 80 mm for the setup presented here. Suppose one period of the interference pattern with original period p must cover at least M pixels of size s on the detector to be analyzed reliably, simple geometric considerations with the assumption $a \ll d$ yield a minimal focal length f_{min}:

$$f_{min} = \frac{(d+a) \cdot p / Ms}{p / Ms - 1} \tag{6.1}$$

With the experimental parameters, this criterion yields $f_{min} = 0.97$ m for the shorter distance 0.2 z_T and $f_{min} = 3$ m for 0.7 z_T.

However, the resolution criterion is not sufficient either, because the focused beam after the lens changes the effective propagation distance L_E between grating and detector and shifts the Talbot distances [128]:

$$L_E = \frac{(f-a) \cdot d}{f-a-d} \tag{6.2}$$

For the distances used in these experiments, the worst case is a difference between L_E and d of 0.3 z_T, because it would place the detector at a distance of 0.5 z_T or z_T, both of which are positions with no intensity contrast, and thus no possibility to analyze the interference pattern. To avoid this and to still retain a high visibility to analyze the interference pattern, we restrict $L_E - d$ to be smaller than $z_T / 8$. Employing this criterion, we obtain a minimum focal length of 1.4 m for 0.2 z_T and 12.5 m for 0.7 z_T. In practice, this means that the interferometer is well suited for the analysis of individual lens elements, which is very helpful to find systematic fabrication errors. It also provides a way to sort out defective elements from a CRL stack to improve the overall lens quality. A fully assembled CRL can only be tested with some limitations: a way to analyze shorter focal lengths with minimal changes to the setup would be to make a negative, i.e. placing the lens between grating and detector to a distance short enough to fulfill both the resolution and the propagation distance criterion. However, this also reduces the sensitivity of the system, because of the short propagation distance between sample and detector [118].

6.2 Data analysis

As described in section 2.2.2, the interferometer measures the differential phase $\frac{\partial\Phi}{\partial x}$, which is proportional to the refraction angle. Figure 6.2 shows the refraction angle for the two orthogonal directions, calculated as described in the previous section, for the example of a beryllium point focus lens. All data analysis in this section will be explained with this lens as example; it has a circular aperture with a diameter of 430 µm, its nominal apex radius is 50 µm. It was scanned at 10 rotational angles in the 0.2 z_T configuration.

Figure 6.2 shows the horizontal and vertical refraction angle. Two concentric rings are visible, one with a diameter of about 480 µm and one with 430 µm; the gradient between the two rings is a lot smaller than inside the inner ring. This is because the lens is fabricated by imprinting from two sides, and the two imprints do not have the same depth. The nominal apex radius of this lens is R = 50 µm, assuming a parabolic profile of the thickness T, $T(x) = 1/2R\, x^2$, the depth can be estimated to 462 µm for the smaller circle and 576 µm for the larger circle, a difference of 114 µm, or about 25 % of the depth of the shallower imprint.

Figure 6.2: Refraction angle (proportional to phase gradient) in orthogonal directions for a beryllium point focus lens with apex radius 50 µm. Adapted from [125]

6.2.1 Local lens aberrations

As the profile of an ideal focusing lens is parabolic, the gradient, and therefore the refraction angle, is linear. The slope $\partial\alpha/\partial x$ is inversely proportional to the apex radius R of the parabola, and consequently to the focal length f:

$$\gamma\frac{\partial\alpha}{\partial x}=\frac{1}{f}=\frac{2\delta}{R} \tag{6.3}$$

Here, a factor γ was introduced to account for the fact that the refraction angle is measured in the detector plane, where the aperture of the lens appears demagnified. The factor can be calculated by determining the diameter l_d of the lens in the detector plane, and afterwards finding the shift of the interference pattern at the edges of the aperture. For a focusing lens, the total shift $\Delta\varphi_t$ between the edges is exactly the difference between the aperture diameter l in the lens plane and the detector plane, $l=l_d+\Delta\phi_t$, which gives the factor γ [125]:

$$\gamma=\frac{l_d}{l_d+\Delta\phi_t} \tag{6.4}$$

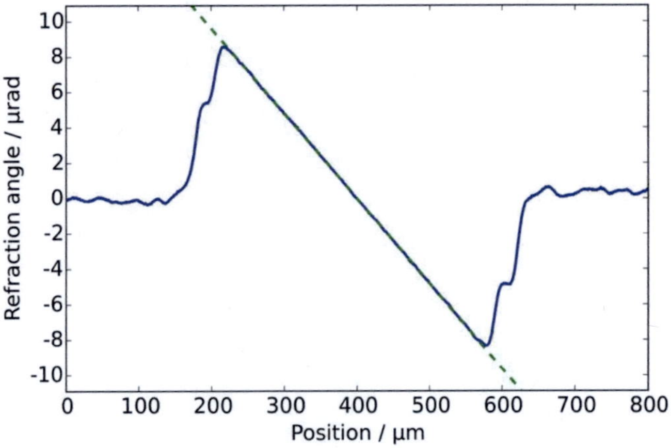

Figure 6.3: Line plot through the center of the beryllium lens in Figure 6.2. The dashed line is a linear fit to the data used to determine the focal length and local aberrations

The apex radius can then be determined from this corrected focal length. Figure 6.3 shows a line plot of the horizontal refraction angle through the center of the beryllium lens in Figure 6.2, and the linear fit to it. The shoulders at the edges of the linear part are again due to the different imprinting depth on the two sides of the lens.

Apart from the determination of the focal length, the linear fit also offers a way to quantify local aberrations of a lens, by subtracting the fit from the data and plotting the residuals. Any non-zero value in the residuals then represents a deviation from the intended refraction angle. The residuals can be plotted in two different ways: either for both orthogonal directions individually (see Figure 6.4), which also shows the sign of the deviation, or in a combined plot, showing only the magnitude (see Figure 6.5). In the plots, distinct, localized points with aberrations in the order of 100 nrad are visible, namely in the center of the aperture, which is most probably caused by deformation of the imprinting tool during the process. A defect of this type has already been reported for another beryllium lens with much larger apex radius [52]. Other defects can be seen off-center, e.g. one ca. 100 µm beneath the center. This type of defect is caused by material inhomogeneity, such as voids and inclusions that are known to appear in the beryllium raw material [121].

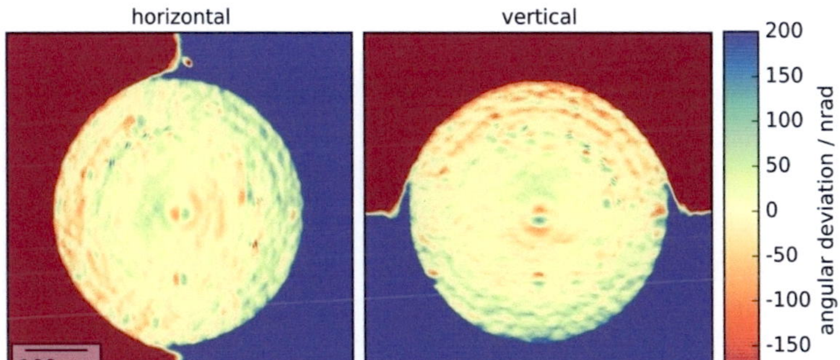

Figure 6.4: Map of the individual components of angular deviation vector for the beryllium lens also presented in Figure 6.2

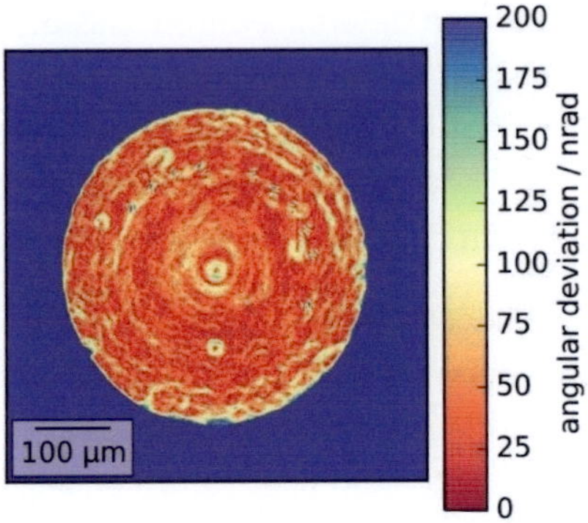

Figure 6.5: Map of the magnitude of angular deviation for the beryllium point focus lens as presented in Figure 6.2. Adapted from [125].

6.2.2 Global lens aberrations

To characterize not only local defects, but lens aberrations like astigmatism or spherical aberration, the phase of the wavefront, rather than its gradient, is needed. This can be determined using 2d Fourier integration, the approach is explained in the appendix [129]. Assuming a homogeneous material, the retrieved wavefront shape is directly proportional to the thickness of the lens, which can be quantitatively retrieved, up to a constant offset, when the refractive index of the material is known. The quantification of shape errors is done by decomposition of the shape into a set of polynomials that must be suited for the particular problem. Spherical apertures with a point focus, like the beryllium lens example, are usually characterized in terms of Zernike polynomials [130]. These polynomials form an orthogonal base of the unit disc and are defined in polar coordinates and characterized by their radial degree n and azimuthal degree m, Z_n^m and can be identified with the common lens aberrations. Any shape S with a circular footprint can be written as $S = \sum_0^\infty A_{nm} Z_n^m$. The orthogonality condition reads as follows, with δ as the Kronecker delta:

$$\int_0^1 \int_0^{2\pi} Z_n^m Z_{n'}^{m'} dr d\phi = \delta_{nn'} \delta_{mm'} \tag{6.5}$$

The strict orthogonality of equation (6.5) only holds true for continuous integration. For discrete datasets, like the image acquired by a camera, there is some crosstalk between coefficients which must be accounted for. Consequently, the decomposition of the lens shape was done stepwise; the contributions with the largest coefficients were subtracted from the data to quantify the aberrations. These contributions were the defocus term, which is the desired shape, tip and tilt, corresponding to misalignment of the axis of the lens to the optical axis of the interferometer, as well as a constant offset with no physical meaning. The coefficients found for the beryllium lens are listed in Table 7, along with the names of the aberrations.

The only notable contributions apart from the defocus term are some astigmatism and horizontal coma, along with the largest contribution of spherical aberration, whose magnitude is still smaller than 0.5 % of the defocus term. From this decomposition, it can be said that the lens is accurate to its shape with a deviation of less than 1 µm. From a plot of the thickness when removing defocus, tip, tilt and the offset, it is also possible to determine whether the local errors of the lens are voids or inclusions, because inclusions would have a higher refractive index decrement and therefore appear as larger thickness than desired, while voids would reduce the thickness. Figure 6.6 shows this plot, and the local errors visible in Figure 6.5 B reappear as dark spots, meaning a lower Be thickness than in the neighboring region, which identifies them as voids in the material. The spherical aberration is also visible as a bright region in the center of the lens. A higher beryllium thickness in the center is consistent with the deformation of the imprinting tool as the cause for the spherical aberration.

Table 7: Zernike coefficients found for the beryllium lens [125]

Name	*n, m*	A_n^m / µm
Defocus	2, 0	-255.7
Oblique astigmatism	**2, -2**	**-0.21**
Vertical astigmatism	2, 2	0.04
Vertical coma	3, -1	-0.10
Horizontal coma	**3, 1**	**0.42**
Vertical trefoil	3, -3	-0.09
Oblique trefoil	3, 3	0.02
Primary spherical	**4, 0**	**0.76**
Vertical secondary astigmatism	4, 2	0.08
Oblique secondary astigmatism	4, -2	-0.08
Vertical quadrafoil	4, 4	-0.01
Oblique quadrafoil	4, -4	-0.05

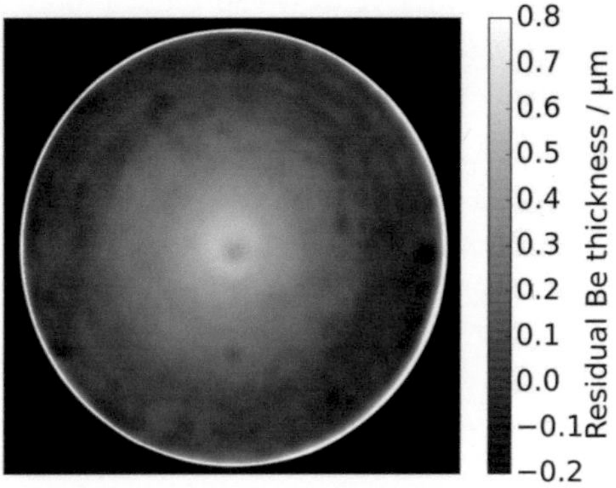

Figure 6.6: Residual thickness of the retrieved beryllium profile after removing defocus, tip, tilt and constant offset

6.3 Analysis of polymer compound refractive lenses

The tool described in the previous section can now be employed to the analysis of CRLs made by X-ray lithography to find the specific types of defects and gain insight to their causes in the fabrication process. Figure 6.7 shows the two orthogonal directions of the refraction angle of a line focus lens element made of mr-X, analyzed in the interferometer configuration with the longer grating-detector distance and scanned at 12 different rotational angles. This element was taken from a CRL intended for the creation of a nanofocus beam [18], the element has a rectangular aperture of 1000 x 800 µm and an apex radius of 300 µm. The two directions for the refraction angle were chosen to represent the focusing and non-focusing direction of the lens and consequently there is no visible vertical gradient in Figure 6.7, only the outline of the aperture is visible due to reflections from the top and the sides of the lens element and imperfect alignment with the optical axis. The stripes on the left edge in the pictures of the line focus lens come from the fact that the lens did not completely fit in the field of view of the interferometer at all rotation angles, and parts of the final image that are not covered by data from all rotation angles are incorrectly normalized. The point focus lens in section 6.2 is much smaller, therefore no such areas are visible there.

Figure 6.7: Refraction angle (proportional to phase gradient) in orthogonal directions for a polymer line focus lens with apex radius 300 µm. Adapted from [125]

The plots of the residuals from the linear fit are shown in Figure 6.8 and Figure 6.9. As the line focus lens should have no gradient in the vertical direction, no linear fit was performed; the map presents the average deviation from the mean value inside the aperture. In contrast to the beryllium lens, no defects of the strongly localized type were found, which is due to the greater homogeneity of the amorphous polymer. Instead, there are stripe-like deviations with a lower spatial frequency that indicate global aberrations. The circular features in the center of the aperture, as well as the speckle-like deviations with high spatial frequency, are not real defects but arise from instabilities in the beam, which is explained in more detail in the appendix. The color scaling in Figure 6.9 is chosen as the same as in the equivalent plot for the beryllium lens in Figure 6.5, to show that the absolute values of deviation are much smaller in the polymer lens, the average deviation in the aperture is only 18 nrad, compared to 47 nrad for the beryllium lens. However, because of the larger differences in aperture and focal length between the two lenses, the absolute values of deviation cannot be directly compared. To achieve the same focal length with the polymer lens as with the beryllium lens, one would need approximately 7 polymer lens elements. Assuming Gaussian error propagation, this would lead to a mean deviation of 48 nrad inside the aperture, which is practically the same as for the single beryllium lens, but on an aperture that is 6.5 times larger.

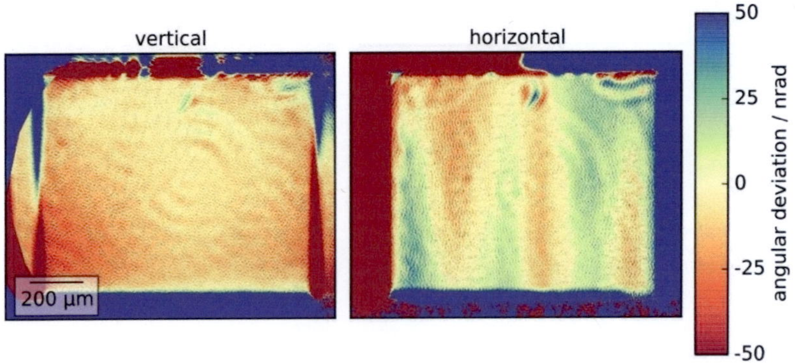

Figure 6.8: Map of the individual components of angular deviation vector for the polymer lens also presented in Figure 6.7

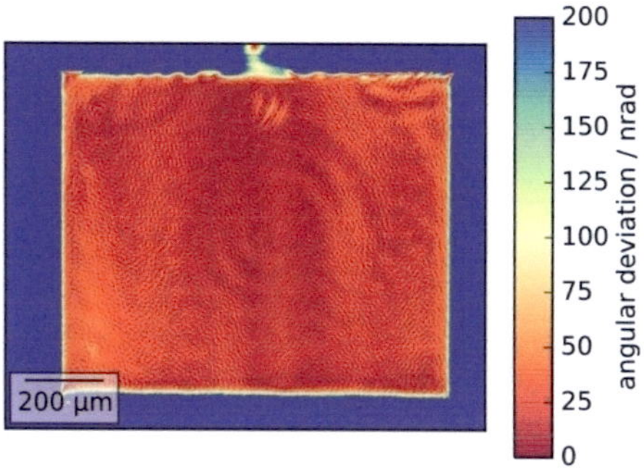

Figure 6.9: Map of the magnitude of angular deviation for the polymer line focus lens as presented in Figure 6.7, the color scaling is identical to the plot for the beryllium lens in Figure 6.5. Adapted from [125].

To quantify global aberrations, the integrated phase is decomposed into a set of polynomials. While the decomposition in Zernike polynomials provides good insight in the properties of the point focus lens, it is not well suited for the characterization of the line focus lens with its rectangular aperture. A set of polynomials must be employed that better matches the symmetry of the problem. In the literature, several possibilities are described, such as Chebyshev or Legendre polynomials, or Zernike polynomials adapted to a rectangular aperture [131]. The adapted Zernike polynomials are not well suited for the problem at hand, because they do not contain contributions that are quadratic in only one dimension, which is the expected main component of the lens. Both Chebyshev and Legendre polynomials have been tried out for the characterization of the retrieved profile of the polymer lens, with the Legendre set yielding much lower residuals after decomposition in the first 15 terms. Thus, the Legendre polynomials suit the symmetry of the problem much better.

The one-dimensional Legendre polynomials of degree n, P_n, are orthogonal on the interval [-1,1]. Constructing the 2d Legendre polynomials L_j is done by simply multiplying the polynomials in x and y:

$$L_j(x,y) = P_n(x) \cdot P_m(y).$$

The orthogonality condition reads as follows, again with δ as the Kronecker Delta:

$$\int_{-1}^{1}\int_{-1}^{1} L_j(x,y)\cdot L_k(x,y)dxdy = 4\cdot\delta_{jk}. \tag{6.6}$$

The decomposition of the polymer lens profile was done in the same iterative way as for the beryllium lens but the crosstalk between coefficients was much less pronounced in this case for two main reasons: On the one hand, the number of pixels inside the aperture is much larger for the polymer lens than for the beryllium lens, providing a more accurate approximation of the integration. On the other hand, the camera pixels have the same symmetry as the rectangular base area of the Legendre polynomials, while the base of the Zernike polynomials is circular. The coefficients found in the decomposition are listed in Table 8. There is no strict association of a specific Legendre polynomial with a lens error as for the Zernike polynomials, but nevertheless a lot of information about these aberrations can be drawn from them. The desired defocus term in this case is described by the coefficient A_6, which describes a focus in the horizontal direction, *y*. Any contribution that is of higher order than 0 in x-direction can be considered a lens error, the largest of which is seen in A_4, which represents focusing in the direction orthogonal to the desired one. This coefficient amounts to about 2 % of the desired coefficient. This effect is undesired and would lead to astigmatism and coma in an assembled point focus lens.

Table 8: 2d-Legendre coefficients found for the polymer lens [125]

j	Polynomial degree in x	Polynomial degree in y	A_j/µm
4	**2**	**0**	**-3.74**
5	1	1	-0.05
6	**0**	**2**	**-188.0**
7	3	0	0.14
8	**2**	**1**	**-1.99**
9	1	2	-1.52
10	0	3	-0.11
11	4	0	-0.12
12	3	1	0.00
13	**2**	**2**	**2.29**
14	1	3	0.02
15	0	4	0.81

Apart from the decomposition into polynomials, it is useful to illustrate the errors of the polymer lens by plotting the measured focal length as a function of structure height above the substrate. This plot can be seen in Figure 6.10, where a continuous, approximately linear increase of the focal length with the height is obvious, the difference between top and bottom is about 2 %. This is consistent with the magnitude of the Legendre coefficients, and shows the same trend as earlier measurements that used the deformation of the projection image of a line grating [132]. The same qualitative behavior was seen in three other polymer lenses with different apex radii, it can therefore be concluded that the effect is systematic to the fabrication technique and not a random defect of this specific lens element.

The cause for the variation of the focal length is not directly revealed by these measurements; possible explanations reach from an effect of the divergence of the beam during exposure, over an effect of beam hardening over the height of the structure to a deformation introduced by the cure shrinkage of the resist material. A simple way to check whether beam divergence or hardening during exposure are responsible for the deviation of the focal length over the height is to check a lens that has been exposed from the backside, through the substrate. As the usual silicon substrate of the lenses would introduce too much absorption and lead to exposure times that are too long for practical application, this is possible only when using a low absorbing substrate, such as the graphite wafers introduced in section 5.2.

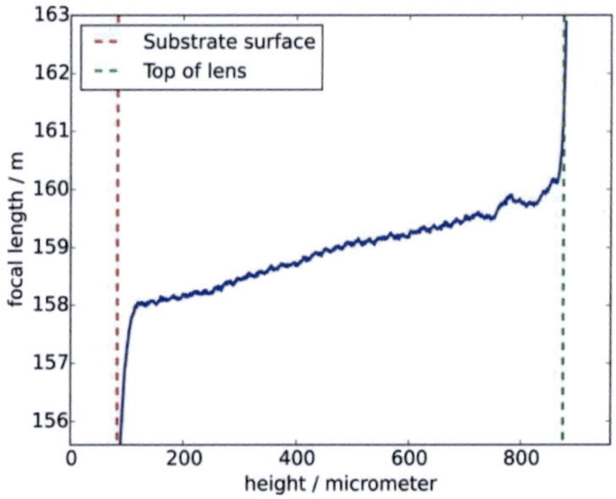

Figure 6.10: Focal length of the polymer focusing lens as function of structure height, adapted from [125]

A lens like this was fabricated and analyzed with the interferometer with the same setup used for the beryllium lens, i.e. at the shorter of the two grating-detector distances, because this lens had a much smaller apex radius than the first polymer lens, and consequently its focal length was only about 5.3 m at 17 keV, which makes it impossible to analyze in the larger Talbot distance (see section 6.1.1). The result for the focal length as a function of structure height can be seen in Figure 6.11. The lens had a structure height of ca. 400 µm. The focal length varies in the same way as for the lenses that were exposed from the front: The focal length at the top is about 2 % larger than close to the substrate. It can therefore be concluded that the gradient in focal length along the height is not caused by effects during the exposure of the photoresist, because these effects would be inverted for a lens that was exposed from the backside.

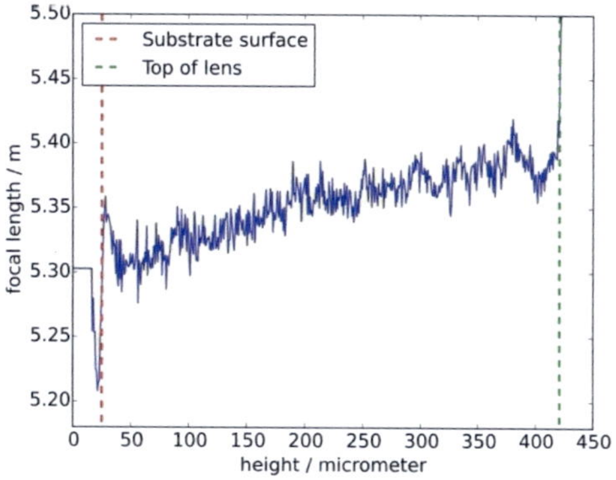

Figure 6.11: Focal length of a second polymer lens fabricated with backside exposure, as a function of structure height

Apart from structure deformation, the variation in focal length could have another cause, which is inhomogeneity of the refractive index of the material. The shorter focal length near the substrate means that the refractive index decrement should be greater near the substrate ground than near the surface. The only obvious explanation for a gradient in the refractive index would be a gradient in the residual solvent content. However, measurements have shown that the residual solvent content should be greater near the substrate surface than near the resist surface [104], from which one would expect a refractive index gradient in the other direction. It is thus likely that the aberrations come from a shape deviation, and the reason must be searched for in the properties of the resist and its behavior and deformation during the process. This happens mainly during the PEB and is therefore closely related to the studies presented in chapter 0. The next section shows an analysis of the deformation using a finite element simulation based on the results of these studies.

6.4 Finite element modeling of photoresist structures

The knowledge of the mechanical properties such as the young's modulus and the curing shrinkage of the resist, whose measurements were described in section 4.4, allows the application of finite element modeling techniques to predict stress and deformation during the fabrication process. Kenntner and Amberger have used this for the consideration of different grating layouts [78], [87]. Apart from this purpose, the procedure can also be used for other lithographic structures like the CRL elements discussed in this chapter.

To investigate the influence of cure shrinkage on a lenses' geometry, a finite element simulation was performed for the geometry of a single lens element. The curing shrinkage was modeled as thermal contraction of the material by 1.5 %, the material was assumed to stick perfectly to the substrate. The value for the shrinkage was assumed lower than what was measured in section 4.4, because not all of the shrinkage introduces stress to the structures. Measurements at other, non-photoactive epoxy systems have shown that the shrinkage roughly follows a linear dependency on the degree of cure, while changes in the mechanical properties only arise at a degree of cure of about 40 % [112], [133]. The simulations were performed using ANSYS Mechanical, the 3d model was constructed from the mask layout with CAD software. The simulation was first performed for the lens with 300 µm apex radius, the height of the model lens was chosen as 800 µm, just as in the experimental data presented in the previous section. Figure 6.12 shows the deformed lens geometry as it is calculated by the FEM simulation, the maximum deformation is about 16 µm, which is very large but mostly to be attributed to a loss in height, other significant deformations affect mostly the edges of the structures.

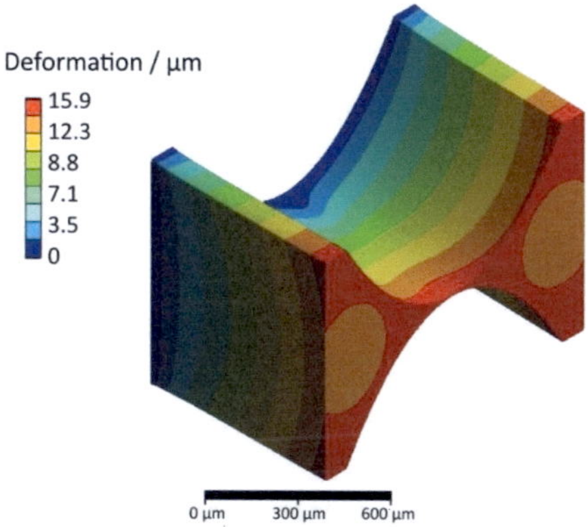

Figure 6.12: Deformed model of a single refractive lens element. The color coding represents the total magnitude of deformation. The blue part of the lens is on the substrate surface, the red part on the top of the lens.

The coordinates of the deformed geometry were exported to a text file and a python script was used to fit a parabolic function to a cross-section of the geometry at both sides at height intervals of 10 µm. The deformed geometry consists of mesh points that are not evenly spaced, the points on the surface of the structure in each height interval were therefore projected to a single line in the middle of this interval and then a fit was performed. The fit function was a second order polynomial of the form $f(x) = a \cdot x^2 + b \cdot x + c$. The parameter a directly relates to the apex radius R of the parabola: $R = 1/2a$ and therefore to the focal length of the lens. The focal length f obtained from the fit, assuming 17 keV photon energy, is plotted in Figure 6.13 as a function of height over the substrate. The dashed blue line represents the undeformed geometry, while the solid green line represents the deformed geometry. The refractive index decrement of the polymer was assumed as $\delta = 9.392 \cdot 10^{-7}$ for 17 keV, following the relation reported by A. Last: $\delta = 2.7142 \cdot 10^{-4} / E^2$, with E in keV [134]. For a quantitative match of the focal length with the results from the interferometer, δ had to be corrected by a factor of 1.02. The total variation of the focal length in the deformed geometry is about 2.5 % of its original value, which is comparable to the experimental findings, but shows a trend in the opposite direction, to shorter focal length near the resist surface.

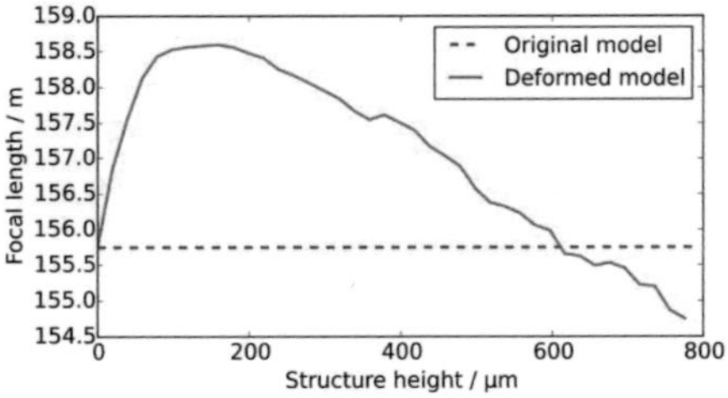

Figure 6.13: Focal length (calculated for 17 keV) of the simulated lens element as calculated from fit as a function of the height over the substrate, assuming a shrinkage of 1.5 %

The model of volumetric shrinkage of the magnitude measured by pycnometry as thermal contraction is therefore not accurate to determine the final shape of microstructures fabricated by X-ray lithography. This is also reflected in the fact that a pure shrinkage cannot account for the waviness found in high aspect ratio grating structures, as described in section 4.4.2. Instead, it turns out that the assumption of a thermal expansion of the theoretical geometry leads to a variation of the focal length of the lens that resembles the experimental results very strongly, which is visible in the plots of Figure 6.14. In both Figure 6.14 A and B, the red, solid line represents the data measured with the interferometer, as presented in Figure 6.10 and Figure 6.11. The green dots mark the focal length as determined from the deformed model, the dashed blue line represents the undeformed theoretical geometry.

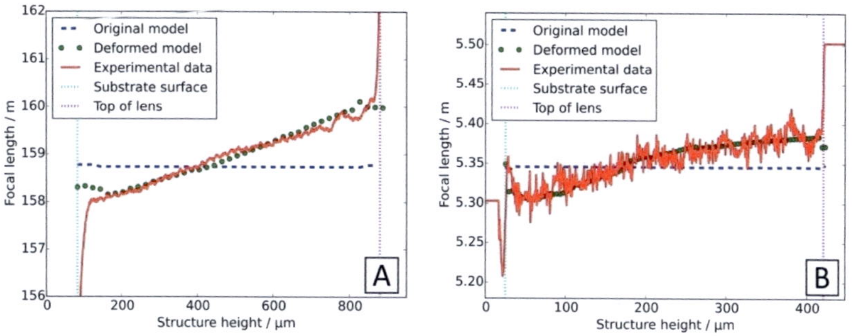

Figure 6.14: Variation of focal length as function of height for two polymer lenses (**A:** f = 159 m, exposure from backside; **B:** f = 5.3 m, standard exposure), experimental and simulated model data. The simulation was performed using an assumption of 1 % expansion for A and 0.6 % expansion for B.

The curves plotted in Figure 6.14 are calculated with an assumption of 1 % expansion for the larger lens with f = 159 m and only 0.6 % for the smaller lens. These curves show excellent agreement with the experimental data. The difference in the assumed expansion for a good fit is an indication that the cause for the deformation is not a pure thermal expansion, but size effects play a role as well. In fact, the evolution of the shape is complicated: After exposure, the resist material is subjected to the PEB, which in the case of the lenses was conducted in the vacuum oven at 75 °C. Due to the vacuum, there is little convection, and the temperature transfer to the material takes place through heat conduction by the substrate and through radiation. The metal shelf of the oven and the silicon substrate are very good heat conductors compared to the polymer; it is thus expected that the temperature of the resist near the substrate surface rises first. Since the heating takes place with the full oven power, and no ramp is used, this results in a temperature gradient throughout the resist at the beginning of PEB. During the rise of the temperature, both substrate and resist expand, but due to the large difference of two orders of magnitude in the thermal expansion coefficients (see section 4.4), the expansion of the substrate can be neglected compared to the uncrosslinked resist. When the temperature surpasses a threshold close to 60 °C, the crosslinking reaction starts, reducing the volume by about 3 % (see section 4.4.1), but due to the temperature gradient, this does not happen simultaneously in the whole resist volume. Additionally, the volume shrinkage is not necessarily isotropic.

These simultaneously acting factors make it very difficult to model the volumetric changes during PEB in the way they are happening, but the very good agreement between simulated and measured shape of the lens elements suggests that thermal expansion provides a good approximation to the shape. Note that this does not mean that the overall resist volume is increasing during the process, because it is not possible to measure a change in height with the interferometer. It is obvious, however, that the cause of the deviation of focal length of the polymer CRLs is to be found in the PEB, and that just as in the fabrication of gratings, the thermal stress is the dominant cause for structural deviation as opposed to stress introduced by chemical shrinkage.

7 Conclusion and Outlook

The main subject of this work is the processing of negative photoresist in X-ray lithography for the fabrication of X-ray gratings which are employed in differential phase contrast imaging (DPCI). High structural quality and optimization of the grating characteristics to the specific DPCI setup are crucial for achieving high image contrast and eventually establishing the technique outside of research laboratories. A series of experiments was conducted to assess the influence of lithography process parameters on the thermomechanical properties of the photoresist material. Tensile testing studies confirmed that the photoresist structures show higher Young's modulus, tensile strength and maximum elongation when increasing the exposure dose. Moreover, the measurements revealed a strong influence of the storage time between coating and exposure of the substrate on its Young's modulus and tensile strength. This was attributed to a change in solvent distribution, and the introduction of a second Prebake step at 40 °C for 24 hours was shown to have the same effect on the mechanical properties as storage for 6 weeks, which makes process management much easier and allows to directly react to new results, e.g. from varied exposure schemes.

Time-resolved gas-pycnometry was used to measure the volumetric shrinkage of the photoresist material during the crosslinking reaction. It was found that with increasing substrate age from one day to 33 days, the time constant of the volume shrinkage dropped from 37 min to 13 min, a decrease of nearly a factor of three. This implies that with fixed postbake parameters, the degree of crosslinking that is achieved in fresh samples is significantly lower than in samples with long storage time, which explains the increase of the Young's modulus and tensile strength with storage time.

Additionally, the cure shrinkage was studied as a function of PEB temperature; the time constant showed a slow decrease with rising temperature, but no difference in the absolute amount of volumetric shrinkage was found, leading to the conclusion that similar mechanical properties can be achieved with lower PEB temperature by increasing the PEB duration. It could be shown that by decreasing the PEB temperature, thus reducing the thermal stress, the structure quality of gratings is increased. The waviness in 800 nm width grating lines with a length of 500 µm and a height of 10 µm could be eliminated by switching from PEB at 75 °C to 60 °C and slowing down the heating process. Both the lower temperature postbake and the second prebake at 40 °C introduced in this work were integrated in the existing process and lead to improvements in the gratings quality. They are expected to benefit the fabrication of other lithographic structures with the mr-X resist as well, also when it is exposed with UV light rather than X-rays.

Apart from the structure quality, the grating substrate plays an important role for imaging. As the radiation is passing through it, it acts as a filter, reducing the intensity especially in the low photon energy range. A loss of photons means a loss of signal-to-noise ratio in the final image, which reduces the sensitivity of a given setup. The silicon wafers used as standard in X-ray lithography have low transmission below 15 keV and negatively affect the performance of grating interferometers working in this range. Therefore, two promising alternatives for grating substrates were introduced in this work, namely graphite and polyimide. Graphite provides the advantage of inherent conductivity, thus eliminating the need for an electroplating seed layer. However, its granular structure and porosity give rise to small angle scattering of X-rays. It is thus not suited as substrate for the phase grating, while it showed good results both as source and analyzer grating. Polyimide substrates can be used for all gratings but require an additional seed layer, which was realized as a Cr/Au layer. Additionally, an adhesion promoter had to be used between seed layer and photoresist to prevent delamination of the resist structure from the substrate.

With a combination of gratings on graphite and polyimide substrates, the sensitivity of a symmetric three grating interferometer could nearly be doubled compared to the previous setup with silicon substrates at constant exposure time. Additionally, the use of low absorbing polyimide substrates enabled the setup of a two-grating interferometer operating at 8.3 keV at the Brazilian Synchrotron laboratory, which would be impossible with standard silicon substrates due to the low transmission. To further increase the sensitivity and enable more compact setups in the future, the introduction of smaller grating periods is necessary.

To see how the resist and processing characteristics influence other X-ray optical components, a set-up was built at the beamline ID06 at the ESRF to test and compare compound refractive X-ray lenses. It is based on grating interferometry with a single grating on a piezo motor and a high-resolution camera, along with a rotation stage for sample mounting. The setup was used to measure the 2d phase gradient introduced by the refractive lens and to quantify aberrations from the desired lens shape in two ways: The residuals of a linear fit to the refraction angle show local lens errors, while the decomposition of the retrieved phase shows global lens aberrations such as spherical aberration or astigmatism.

This methodology was used to analyze the different defects and aberrations of line focus X-ray lenses made from mr-X resist by the same lithographic process used for grating fabrication and to compare them to beryllium lenses made by imprinting. It was found that the lithographically produced lenses had a superior material homogeneity and did not show strong local deviations, while in beryllium voids and inclusions lead to localized errors. On the other hand, the overall shape accuracy appeared higher in the imprinted beryllium lenses, with only little spherical aberration and coma found. The focal length of the line focus polymer lenses was found to increase by about 2 % from substrate ground to the top of the structures, possibly leading to astigmatism and coma in an assembled point focus lens.

The investigation of polymer lenses with different focal length all showed the same trend in the focal length. Effects from the lithographic exposure could be excluded as the reason when the same behavior was also found in a lens exposed from the backside. Using FEM simulation, with the mechanical properties measured by tensile testing as input parameters, the behavior could be accurately modeled by a thermal expansion of the ideal shape of the lens. This implies that the shape deviation is caused by thermal stress and most likely has the same reason as the waviness of long lamella grating structures, which means that the modified process conditions that showed good results in grating fabrication also promise to enable the fabrication of X-ray lenses with significantly less aberrations.

These results highlight the importance seemingly simple process parameters can have for the final structure quality achieved in X-ray lithography. Future investigations might be carried out to identify the role of the different resist components at each processing step; this might also provide a key to explain the lack of reproducibility of microstructure quality between different resist lots.

Appendix

A Phase retrieval

The phase signal recorded by a Talbot interferometer is the derivative $\partial\Phi(x,y)/\partial x$ of the phase in the direction perpendicular to the grating lines. For some applications it is interesting to know the actual shape of the wave front $\Phi(x,y)$. This can be achieved, apart from a constant offset, by simple one-dimensional integration of the signal. However, a problem of this method is that statistical errors in one pixel propagate along the integration direction, leading to stripe artefacts in the retrieved image. A way around this is to measure the phase gradient in two dimensions and apply a procedure based on the Fourier integration theorem. The idea is to combine the two perpendicular gradients in a complex field $g(x,y)$ as [129]

$$g(x,y) = \frac{\partial\Phi(x,y)}{\partial x} + i\frac{\partial\Phi(x,y)}{\partial y}.$$

Its Fourier transform can then be written as

$$\mathcal{F}\left[g(x,y)\right](k,l) = \iint_{-\infty}^{\infty} g(x,y)\exp\left(-2\pi(ikx+ly)\right)dxdy = 2\pi i(k+il)\cdot\mathcal{F}\left[\Phi(x,y)\right](k,l). \tag{8.1}$$

The phase of the wave front can thus be written as an inverse Fourier transform

$$\Phi(x,y) = \mathcal{F}^{-1}\left[\frac{\mathcal{F}\left[g(x,y)\right](k,l)}{2\pi i(k+il)}\right](x,y). \tag{8.2}$$

Integration of the differential phase in this way yields an image with far less artefacts.

B Artefacts in angular deviation maps

As pointed out in chapter 0, each lens was scanned at several azimuthal angles to oversample the two-dimensional phase gradient. This can be used to identify artefacts in the final images. In the polymer lens, concentric circular features reminiscent of ring artifacts in tomography are visible in the angular deviation map. There is no step in the lithographic process that would be likely to cause such a deviation and if one looks at the individual images at different rotation angles, stripe artefacts can be seen that stem from the grating lines, see Figure 7.1 for an example.

Figure 7.1: Processed phase gradient for a single rotation angle of a polymer line focus lens. Horizontal stripe artefacts are visible

Through the superposition of many images with these stripe artefacts, a spherical feature appears, with the axis of rotation of the lens as the center. When using only two azimuthal angles for the calculation of the angular deviation map, the spherical feature becomes rectangular, which clearly shows that it is an artefact and not a true deviation of the lens. The same is true for the high spatial-frequency jitter in the angular deviation maps, which also changes shape when the number of azimuthal scans is changed. The reason for the stripe artifacts must be either a mechanical instability of grating or detector, or a movement of the beam, which can be caused by the monochromator.

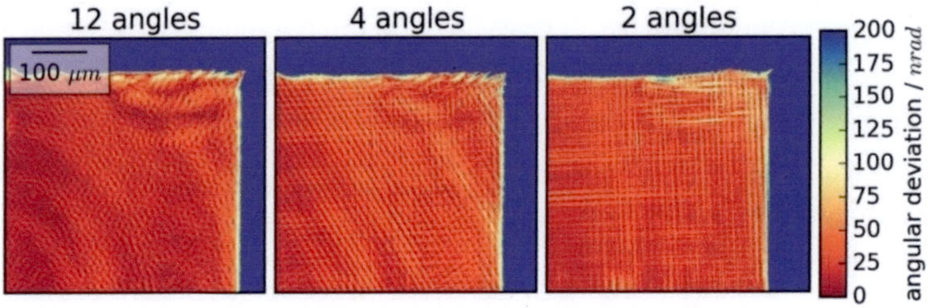

Figure 7.2: Part of the angular deviation map of the polymer line focus lens, calculated using scans at 12, 4 and 2 different azimuthal angles, from [125]

A second type of artefact appears when a grating defect is located inside the aperture of a lens to be analyzed. The single grating, high resolution interferometer is especially vulnerable to phase grating defects because any defect that is larger than one pixel will appear at least partly on another pixel set when the beam is deflected by a sample and therefore will not be corrected by the flat field scan. It will then appear in the phase gradient images both at the original and at the shifted position. An example for such a defect is shown in Figure 7.3, a magnification of the angular deviation map for the beryllium point focus lens presented in section 6.2, again calculated with a different number of azimuthal angles. The defect appears four times in the magnified region of the image reconstructed with 10 angles, two times when using four angles and one time when using two angles, because the defect is at a different position inside the lens aperture for each scan. With increased number of angles, the defect multiplies but is leveled out by the contributions from other angles.

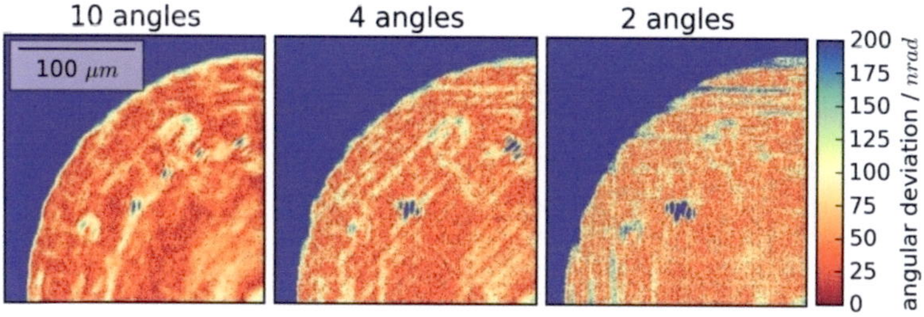

Figure 7.3: Part of the angular deviation map of the beryllium point focus lens, calculated using scans at 10, 4 and 2 different azimuthal angles, from [125]

An error in the determination of the refraction angle will also affect the integrated phase and might therefore affect the results of the decomposition of the retrieved wavefront shape into Zernike or Legendre polynomials. To test this influence, both the concentric circular features and the localized feature from the grating defect were modeled, integrated and decomposed in the same way as the experimental data, but no defect yielded a coefficient larger than 0.03 µm, which is significantly smaller than the aberrations found and therefore does not need to be taken into account for the characterization of global lens errors.

C List of own publications

Peer-reviewed Publication list (in chronological order)

D. Kunka, J. Mohr, V. Nazmov, J. Meiser, P. Meyer, F. Koch, M. Amberger, J. Schulz, M. Walter, T. Duttenhofer, A. Voigt, G. Ahrens, G. Grützner, "Characterization method for new resist formulations for HAR patterns made by X-ray lithography," Microsyst. Technol. (2013). doi:10.1007/s00542-013-2055-x.

M. Ruiz-Yaniz, F. Koch, I. Zanette, A. Rack, P. Meyer, D. Kunka, A. Hipp, J. Mohr, F. Pfeiffer, "X-ray grating interferometry at photon energies over 180 keV," Appl. Phys. Lett. 106 (2015) 151105. doi:10.1063/1.4917293.

F. Koch, F. Marschall, J. Meiser, O. Márkus, A. Faisal, T. Schröter, P. Meyer, D. Kunka, A. Last, J. Mohr, "Increasing the aperture of x-ray mosaic lenses by freeze drying," J. Micromechanics Microengineering. 25 (2015) 75015. doi:10.1088/0960-1317/25/7/075015.

F.J. Koch, T.J. Schröter, D. Kunka, P. Meyer, J. Meiser, A. Faisal, M.I. Khalil, L. Birnbacher, M. Viermetz, M. Walter, J. Schulz, F. Pfeiffer, J. Mohr, "Note: Gratings on low absorbing substrates for x-ray phase contrast imaging," Rev. Sci. Instrum. 86 (2015) 126114. doi:10.1063/1.4939055.

L. Birnbacher, M. Willner, A. Velroyen, M. Marschner, A. Hipp, J. Meiser, F. Koch, T. Schröter, D. Kunka, J. Mohr, F. Pfeiffer, J. Herzen, "Experimental Realisation of High-sensitivity Laboratory X-ray Grating-based Phase-contrast Computed Tomography," Sci. Rep. 6 (2016) 24022. doi:10.1038/srep24022.

F.J. Koch, C. Detlefs, T.J. Schröter, D. Kunka, A. Last, J. Mohr, "Quantitative characterization of X-ray lenses from two fabrication techniques with grating interferometry," Opt. Express. 24 (2016) 9168. doi:10.1364/OE.24.009168.

J. Meiser, M. Willner, T. Schröter, A. Hofmann, J. Rieger, F. Koch, L. Birnbacher, M. Schüttler, D. Kunka, P. Meyer, A. Faisal, M. Amberger,

T. Duttenhofer, T. Weber, A. Hipp, S. Ehn, M. Walter, J. Herzen, J. Schulz, F. Pfeiffer, J. Mohr, "Increasing the field of view in grating based X-ray phase contrast imaging using stitched gratings," J. Xray. Sci. Technol. 24 (2016) 379–388. doi:10.3233/XST-160552.

T. J. Schröter, F. Koch, P. Meyer, M. Baumann, D. Münch, D. Kunka, S. Engelhardt, M. Zuber, T. Baumbach, and J. Mohr, "Large area gratings by x-ray LIGA dynamic exposure for x-ray phase-contrast imaging," *J. Micro/Nanolithography, MEMS, MOEMS*, vol. 16, no. 1, pp. 013501–013501, 2017.

T. J. Schröter, F. J. Koch, P. Meyer, D. Kunka, J. Meiser, K. Willer, L. Gromann, F. D. Marco, J. Herzen, P. Noel, A. Yaroshenko, A. Hofmann, F. Pfeiffer, and J. Mohr, "Large field-of-view tiled grating structures for X-ray phase-contrast imaging," *Rev. Sci. Instrum.*, vol. 88, no. 1, p. 15104, Jan. 2017.

F.J. Koch, F.P. O'Dowd, M.B. Cardoso, R.R. Da Silva, M. Cavicchioli, S.J.L. Ribeiro, T.J. Schröter, A. Faisal, P. Meyer, D. Kunka, J. Mohr, "Low energy X-ray grating interferometry at the Brazilian Synchrotron," Opt. Commun. 393 (2017) 195–198. doi:10.1016/j.optcom.2017.02.055.

Conference Contributions (only as presenting author)

F. Koch, T. Schröter, A. Faisal, J. Meiser, P. Meyer, D. Kunka, A. Last, I. Zanette, A. Rack, C. Detlefs and J. Mohr, "X-ray Grating Interferometer for Non-Destructive Testing and Quantitative Characterization of Refractive X-ray Optics;" Oral presentation, XNPIG 2015, Bethesda, USA

F. Koch, T. Schröter, J. Meiser, P. Meyer, D. Kunka, A. Faisal, A. Yaroshenko, L. Birnbacher, F. Prade, F. Pfeiffer, T. Duttenhofer, J. Schulz, J. Mohr, „Mit Röntgenlithographie gefertigte Gitterstrukturen ermöglichen neue Röntgenbildgebung in Medizin und Materialwissenschaft;" Oral presentation, Mikrosystemtechnik Kongress 2015, VDE Verlag, Karlsruhe, Germany

F. Koch, J. Mohr, D. Kunka, P. Meyer, J. Meiser, M. Amberger, M. Börner, M. Walter, J. Schulz, "LIGA-Gratings for DPCI – The state of the art;" Poster presentation, DPG spring meeting 2014, Mainz, Germany

J. Meiser, F. Koch, T. Schröter, D. Kunka, L. Birnbacher, I. Khalil, T. Duttenhofer, P. Meyer, M. Walter, F. Pfeiffer and J. Mohr, "Dose reduction using low absorbing grating substrates for biomedical imaging at energies <30 keV;" Poster presentation, IMXP 2015, Garmisch-Partenkirchen, Germany

F.J. Koch, C. Detlefs, T.J. Schröter, O. Markus, D. Kunka, A. Last, J. Mohr, "Comparison of X-ray lenses using grating interferometry;" Poster presentation, XRM 2016, Oxford, UK

References

[1] W. C. Röntgen, "Über eine neue Art von Strahlen. (Vorläufige Mittheilung.)," *Sitzungsberichte der Würzburg. Phys. Gesellschaft*, 1895.

[2] J. T. Bushberg, J. A. Seibert, E. M. Leidholt, and J. M. Boone, *The Essential Physics of Medical Imaging*, 3rd ed. Philadelphia: Lippincott Williams & Wolters Kluwer, 2012.

[3] F. Smith, *Industrial applications of X-ray diffraction*. CRC Press, 1999.

[4] D. C. Koningsberger and R. Prins, "X-ray absorption: principles, applications, techniques of EXAFS, SEXAFS, and XANES." John Wiley and Sons,New York, NY, 1988.

[5] R. F. Mould, *A century of x-rays and radioactivity in medicine : with emphasis on photographic records of the early years*. Bristol; Philadelphia: Institute of Physics Publishing, 1993.

[6] A. Momose, T. Takeda, A. Yoneyama, I. Koyama, and Y. Itai, "Phase-Contrast X-Ray Imaging Using an X-Ray Interferometer for Biological Imaging," *Anal. Sci.*, vol. 17, pp. 1–4, 2001.

[7] A. Bravin, P. Coan, and P. Suortti, "X-ray phase-contrast imaging: from pre-clinical applications towards clinics.," *Phys. Med. Biol.*, vol. 58, no. 1, pp. R1-35, Jan. 2013.

[8] A. Momose, "Phase-sensitive imaging and phase tomography using X-ray interferometers.," *Opt. Express*, vol. 11, pp. 2303–2314, 2003.

[9] T. Weitkamp, A. Diaz, C. David, F. Pfeiffer, M. Stampanoni, P. Cloetens, and E. Ziegler, "X-ray phase imaging with a grating interferometer.," *Opt. Express*, vol. 13, pp. 6296–6304, 2005.

[10] F. Pfeiffer, T. Weitkamp, O. Bunk, and C. David, “Phase retrieval and differential phase-contrast imaging with low-brilliance X-ray sources,” *Nat. Phys.*, vol. 2, no. 4, pp. 258–261, Mar. 2006.

[11] H. F. Talbot, “LXXVI. Facts relating to optical science. No. IV,” ... *Philos. Mag. J. Sci.*, vol. 9, no. 56, pp. 401–407, Dec. 1836.

[12] F. Pfeiffer, M. Bech, O. Bunk, P. Kraft, E. F. Eikenberry, C. Brönnimann, C. Grünzweig, and C. David, “Hard-X-ray dark-field imaging using a grating interferometer.,” *Nat. Mater.*, vol. 7, pp. 134–137, 2008.

[13] E. W. Becker, W. Ehrfeld, P. Hagmann, A. Maner, and D. Münchmeyer, “Fabrication of microstructures with high aspect ratios and great structural heights by synchrotron radiation lithography, galvanoforming, and plastic moulding (LIGA process),” *Microelectronic Engineering*, vol. 4. pp. 35–56, 1986.

[14] E. Reznikova, J. Mohr, M. Boerner, J. Mathuni, V. Nazmov, B. Matthis, and P.-J. Jakobs, “Fabrication of high aspect ratio submicron gratings by soft X-ray SU-8 lithography,” *Microsyst. Technol.*, 2008.

[15] H. Lorenz, M. Despont, N. Fahrni, N. LaBianca, P. Renaud, and P. Vettiger, “SU-8: a low-cost negative resist for MEMS,” *Journal of Micromechanics and Microengineering*, vol. 7, no. 3. pp. 121–124, 1999.

[16] A. Snigirev, V. Kohn, I. Snigireva, and B. Lengeler, “A compound refractive lens for focusing high-energy X-rays,” *Nature*, vol. 384, no. 6604. pp. 49–51, 1996.

[17] V. Nazmov, E. Reznikova, J. Mohr, A. Snigirev, I. Snigireva, S. Achenbach, and V. Saile, “Fabrication and preliminary testing of X-ray lenses in thick SU-8 resist layers,” *Microsyst. Technol.*, vol. 10, no. 10, pp. 716–721, Dec. 2004.

[18] C. Krywka, A. Last, F. Marschall, O. Márkus, S. Georgi, M. Müller, J. Mohr, C. Krywka, A. Last, F. Marschall, O. Márkus, S. Georgi, M. Müller,

and J. Mohr, "Polymer compound refractive lenses for hard X-ray nanofocusing Polymer Compound Refractive Lenses for Hard X-ray Nanofocusing," in *AIP Conference Proceedings*, 2016, vol. 1764.

[19] M. Lyubomirskiy, I. Snigireva, V. Kohn, S. Kuznetsov, V. Yunkin, G. Vaughan, and A. Snigirev, "30-Lens interferometer for high-energy X-rays," *J. Synchrotron Radiat.*, vol. 23, no. 5, pp. 1–6, 2016.

[20] J. Hilhorst, F. Marschall, T. N. Tran Thi, A. Last, and T. Schülli, "Full-field X-ray diffraction microscopy using polymeric compound refractive lenses," *J. Appl. Crystallogr.*, vol. 47, no. 6, pp. 1882–1888, 2014.

[21] R. Behling, *Modern diagnostic x-ray sources : technology, manufacturing, reliability*. CRC Press, 2015.

[22] P. Frame, "Rotating Anode X-ray Tubes." [Online]. Available: https://www.orau.org/ptp/collection/xraytubescoolidge/rotating anode information.htm. [Accessed: 07-Oct-2016].

[23] "Siemens: X-ray Tubes & X-ray Tube Assemblies." [Online]. Available: https://www.oem-products.siemens.com/x-ray-tube. [Accessed: 07 Oct 2016].

[24] M. Otendal, T. Tuohimaa, U. Vogt, and H. M. Hertz, "A 9 keV electron-impact liquid-gallium-jet x-ray source.," *Rev. Sci. Instrum.*, vol. 79, no. 1, p. 16102, Jan. 2008.

[25] K. Wille, *The Physics of Particle Accelerators*. Oxford, UK: Oxford University Press, 2000.

[26] J. Als-Nielsen and D. McMorrow, *Elements of Modern X-Ray Physics*, 2nd ed. John Wiley & Sons, 2011.

[27] A. Singer, I. A. Vartanyants, M. Kuhlmann, S. Duesterer, R. Treusch, and J. Feldhaus, "Transverse-Coherence Properties of the Free-Electron-Laser FLASH at DESY," *Phys. Rev. Lett.*, vol. 101, no. 25, p. 254801, Dec. 2008.

[28] D. Paganin, *Coherent X-Ray Optics*. Oxford University Press, 2006.

[29] M. Born and E. Wolf, *Principles of optics*, 7th ed. Cambridge: Cambridge University Press, 2009.

[30] A.-J. Fresnel, "Calcul de l'intensite de la lumiere au centre de l'ombre d'un ecran," *Oeuvres Complet. d' Augustin Fresnel*, vol. 1, 1866.

[31] "Fresnel zone plate theory, generation, tolerancing, fabrication, and applications.," *The Center For X-Ray Optics Lawrence Berkeley National Laboratory*. [Online]. Available: http://zoneplate.lbl.gov/. [Accessed: 01 May 2016].

[32] W. Chao, J. Kim, S. Rekawa, P. Fischer, and E. H. Anderson, "Demonstration of 12 nm resolution Fresnel zone plate lens based soft x-ray microscopy.," *Opt. Express*, vol. 17, no. 20, pp. 17669–77, Sep. 2009.

[33] P. Kirkpatrick and A. V. Baez, "Formation of Optical Images by X-Rays," *J. Opt. Soc. Am.*, vol. 38, no. 9, p. 766, Sep. 1948.

[34] A. Last, "X-ray optics." [Online]. Available: www.x-ray-optics.de. [Accessed: 20 Jun 2005].

[35] B. Lengeler, C. Schroer, J. Tümmler, B. Benner, M. Richwin, A. Snigirev, I. Snigireva, and M. Drakopoulos, "Imaging by parabolic refractive lenses in the hard X-ray range," *J. Synchrotron Radiat.*, vol. 6, no. 6, pp. 1153–1167, Nov. 1999.

[36] "RXOPTICS." [Online]. Available: www.rxoptics.de. [Accessed: 01 May 2016].

[37] V. Aristov, M. Grigoriev, S. Kuznetsov, L. Shabelnikov, V. Yunkin, T. Weitkamp, C. Rau, I. Snigireva, A. Snigirev, M. Hoffmann, and E. Voges, "X-ray refractive planar lens with minimized absorption," *Appl. Phys. Lett.*, vol. 77, no. 24, p. 4058, Dec. 2000.

[38] U. Bonse and M. Hart, "An X-ray interferometer," *Appl. Phys. Lett.*, vol. 6, no. 8, p. 155, Nov. 1965.

[39] T. J. Davis, D. Gao, T. E. Gureyev, A. W. Stevenson, and S. W. Wilkins, "Phase-contrast imaging of weakly absorbing materials using hard X-rays," *Nature*, vol. 373. pp. 595–598, 1995.

[40] A. Snigirev, I. Snigireva, V. Kohn, S. Kuznetsov, and I. Schelokov, "On the possibilities of x-ray phase contrast microimaging by coherent high-energy synchrotron radiation," *Rev. Sci. Instrum.*, vol. 66, no. 12, p. 5486, Dec. 1995.

[41] P. Cloetens, W. Ludwig, J. Baruchel, D. Van Dyck, J. Van Landuyt, J. P. Guigay, and M. Schlenker, "Holotomography: Quantitative phase tomography with micrometer resolution using hard synchrotron radiation x rays," *Appl. Phys. Lett.*, vol. 75, no. 19, p. 2912, 1999.

[42] I. Zanette, T. Zhou, A. Burvall, U. Lundström, D. H. Larsson, M. Zdora, P. Thibault, F. Pfeiffer, and H. M. Hertz, "Speckle-Based X-Ray Phase-Contrast and Dark-Field Imaging with a Laboratory Source," vol. 253903, no. June, pp. 1–5, 2014.

[43] M. Engelhardt, C. Kottler, O. Bunk, C. David, C. Schroer, J. Baumann, M. Schuster, and F. Pfeiffer, "The fractional Talbot effect in differential x-ray phase-contrast imaging for extended and polychromatic x-ray sources.," *J. Microsc.*, vol. 232, no. 1, pp. 145–57, Oct. 2008.

[44] F. Koch, T. Schröter, J. Meiser, P. Meyer, D. Kunka, A. Faisal, A. Yaroshenko, L. Birnbacher, F. Prade, F. Pfeiffer, T. Duttenhofer, J. Schulz, and J. Mohr, "Mit Röntgenlithographie gefertigte Gitterstrukturen ermöglichen neue Röntgenbildgebung in Medizin und Materialwissenschaft," in *Mikrosystemtechnik Kongress*, 2015.

[45] F. Pfeiffer, C. Grünzweig, O. Bunk, G. Frei, E. Lehmann, and C. David, "Neutron phase imaging and tomography," *Phys. Rev. Lett.*, vol. 96, no. 21, pp. 1–4, 2006.

[46] T. Weitkamp, C. David, C. Kottler, O. Bunk, and F. Pfeiffer, "Tomography with grating interferometers at low-brilliance sources," in *SPIE Optics + Photonics*, 2006, p. 63180S–63180S–10.

[47] A. Malecki, "JGratingCalc." [Online]. Available: http://www.e17.ph.tum.de/index.php/services/software/110.html.

[48] T. Thüring, S. Hämmerle, S. Weiss, J. Nüesch, J. Meiser, J. Mohr, C. David, and M. Stampanoni, "Compact hard X-ray grating interferometry for table top phase contrast micro CT," in *SPIE Medical Imaging*, 2013, p. 866813.

[49] M. Willner, M. Bech, J. Herzen, I. Zanette, D. Hahn, J. Kenntner, J. Mohr, A. Rack, T. Weitkamp, and F. Pfeiffer, "Quantitative X-ray phase-contrast computed tomography at 82 keV.," *Opt. Express*, vol. 21, no. 4, pp. 4155–66, 2013.

[50] A. Momose, W. Yashiro, K. Kido, J. Kiyohara, C. Makifuchi, T. Ito, S. Nagatsuka, C. Honda, D. Noda, T. Hattori, T. Endo, M. Nagashima, and J. Tanaka, "X-ray phase imaging: from synchrotron to hospital.," *Philos. Trans. A. Math. Phys. Eng. Sci.*, vol. 372, no. 2010, p. 20130023, Mar. 2014.

[51] T. Weitkamp, B. Nöhammer, A. Diaz, C. David, and E. Ziegler, "X-ray wavefront analysis and optics characterization with a grating interferometer," *Appl. Phys. Lett.*, vol. 86, no. 5, p. 54101, Jan. 2005.

[52] S. Rutishauser, I. Zanette, T. Weitkamp, T. Donath, and C. David, "At-wavelength characterization of refractive x-ray lenses using a two-dimensional grating interferometer," *Appl. Phys. Lett.*, vol. 99, no. 22, p. 221104, Nov. 2011.

[53] A. Yaroshenko, F. G. Meinel, M. Bech, A. Tapfer, A. Velroyen, S. Schleede, S. D. Auweter, A. Bohla, A. Ö. Yildirim, K. Nikolaou, F. Bamberg, O. Eickelberg, M. F. Reiser, and F. Pfeiffer, "Pulmonary emphysema diagnosis with a preclinical small-animal X-ray dark-field scatter-contrast scanner.," *Radiology*, vol. 269, no. 2, pp. 427–33, Nov. 2013.

[54] T. Michel, J. Rieger, G. Anton, F. Bayer, M. W. Beckmann, J. Durst, P. A. Fasching, W. Haas, A. Hartmann, G. Pelzer, M. Radicke, C. Rauh, A. Ritter, P. Sievers, R. Schulz-Wendtland, M. Uder, D. L. Wachter, T. Weber, E. Wenkel, and A. Zang, "On a dark-field signal generated by micrometer-sized calcifications in phase-contrast mammography.," *Phys. Med. Biol.*, vol. 58, pp. 2713–32, 2013.

[55] S. D. Auweter, J. Herzen, M. Willner, S. Grandl, K. Scherer, F. Bamberg, M. F. Reiser, F. Pfeiffer, and K. Hellerhoff, "X-ray phase-contrast imaging of the breast--advances towards clinical implementation.," *Br. J. Radiol.*, vol. 87, no. 1034, p. 20130606, Feb. 2014.

[56] K. Scherer, E. Braig, K. Willer, M. Willner, A. A. Fingerle, M. Chabior, J. Herzen, M. Eiber, B. Haller, M. Straub, H. Schneider, E. J. Rummeny, P. B. Noël, and F. Pfeiffer, "Non-invasive differentiation of kidney stone types using X-ray dark-field radiography.," *Sci. Rep.*, vol. 5, p. 9527, Jan. 2015.

[57] F. Yang, F. Prade, M. Griffa, I. Jerjen, C. Di Bella, J. Herzen, A. Sarapata, F. Pfeiffer, and P. Lura, "Dark-field X-ray imaging of unsaturated water transport in porous materials," *Appl. Phys. Lett.*, vol. 105, no. 15, p. 154105, Oct. 2014.

[58] A. Malecki, G. Potdevin, T. Biernath, E. Eggl, K. Willer, T. Lasser, J. Maisenbacher, J. Gibmeier, A. Wanner, and F. Pfeiffer, "X-ray tensor tomography," *EPL (Europhysics Lett.*, vol. 105, no. 3, p. 38002, Feb. 2014.

[59] C. Hannesschläger, V. Revol, B. Plank, D. Salaberger, and J. Kastner, "Fibre structure characterisation of injection moulded short fibre-reinforced polymers by X-ray scatter dark field tomography," *Case Stud. Nondestruct. Test. Eval.*, vol. 3, pp. 34–41, Apr. 2015.

[60] A. Tapfer, M. Bech, A. Velroyen, J. Meiser, J. Mohr, M. Walter, J. Schulz, B. Pauwels, P. Bruyndonckx, X. Liu, A. Sasov, and F. Pfeiffer, "Experimental results from a preclinical X-ray phase-contrast CT scanner.," *Proc. Natl. Acad. Sci. U. S. A.*, vol. 109, no. 39, pp. 15691–6, Sep. 2012.

[61] C. T. Chantler, K. Olsen, R. A. Dragoset, J. Chang, A. R. Kishore, S. A. Kotochigova, and D. S. Zucker, "NIST X-Ray Form Factor, Atten., and Scattering Database," *NIST Stand. Ref. Database 66*, 2005.

[62] C. David, J. Bruder, T. Rohbeck, C. Grünzweig, C. Kottler, A. Diaz, O. Bunk, and F. Pfeiffer, "Fabrication of diffraction gratings for hard X-ray phase contrast imaging," *Microelectron. Eng.*, vol. 84, pp. 1172-1177, 2007.

[63] W. Yashiro, D. Noda, T. Hattori, K. Hayashi, A. Momose, and H. Kato, "A metallic glass grating for X-ray grating interferometers fabricated by imprinting," *Appl. Phys. Express*, vol. 7, no. 3, p. 32501, Mar. 2014.

[64] S. K. Lynch, C. Liu, N. Y. Morgan, X. Xiao, A. A. Gomella, D. Mazilu, E. E. Bennett, L. Assoufid, F. de Carlo, and H. Wen, "Fabrication of 200 nm period centimeter area hard x-ray absorption gratings by multilayer deposition," *J. Micromechanics Microengineering*, vol. 105007, no. 22, 2012.

[65] C. Chang and A. Sakdinawat, "Ultra-high aspect ratio high-resolution nanofabrication for hard X-ray diffractive optics.," *Nat. Commun.*, vol. 5, p. 4243, Jan. 2014.

[66] C. Vieu, F. Carcenac, A. Pépin, Y. Chen, M. Mejias, A. Lebib, L. Manin-Ferlazzo, L. Couraud, and H. Launois, "Electron beam lithography: resolution limits and applications," *Appl. Surf. Sci.*, vol. 164, no. 1–4, pp. 111–117, Sep. 2000.

[67] S. Maruo, O. Nakamura, and S. Kawata, "Three-dimensional microfabrication with two-photon-absorbed photopolymerization," *Opt. Lett.*, vol. 22, no. 2, p. 132, Jan. 1997.

[68] S. Y. Chou, "Nanoimprint lithography," *J. Vac. Sci. Technol. B Microelectron. Nanom. Struct.*, vol. 14, no. 6, p. 4129, Nov. 1996.

[69] E. Reznikova, J. Mohr, M. Boerner, V. Nazmov, and P.-J. Jakobs, "Soft X-ray lithography of high aspect ratio SU8 submicron structures," *Microsyst. Technol.*, vol. 14, no. 9–11, pp. 1683–1688, Jan. 2008.

[70] T. J. Schröter, F. J. Koch, P. Meyer, D. Kunka, J. Meiser, K. Willer, L. Gromann, F. D. Marco, J. Herzen, P. Noel, A. Yaroshenko, A. Hofmann, F. Pfeiffer, and J. Mohr, "Large field-of-view tiled grating structures for X-ray phase-contrast imaging," *Rev. Sci. Instrum.*, vol. 88, no. 1, p. 15104, Jan. 2017.

[71] T. J. Schröter, F. Koch, P. Meyer, M. Baumann, D. Münch, D. Kunka, S. Engelhardt, M. Zuber, T. Baumbach, and J. Mohr, "Large area gratings by x-ray LIGA dynamic exposure for x-ray phase-contrast imaging," *J. Micro/Nanolithography, MEMS, MOEMS*, vol. 16, no. 1, pp. 013501–013501, 2017.

[72] P. Meyer, "Fast and accurate X-ray lithography simulation enabled by using Monte Carlo method. New version of DoseSim: a software dedicated to deep X-ray lithography (LIGA)," *Microsyst. Technol.*, vol. 18, no. 12, pp. 1971–1980, Aug. 2012.

[73] P. Atkins and J. De Paula, *Elements of Physical Chemistry*, 4th ed. Oxford, UK: Oxford University Press, 2005.

[74] R. H. Stokes and R. P. Tomlins, "Thermodynamic functions of melting for cyclohexane," *J. Chem. Thermodyn.*, vol. 6, no. 4, pp. 379–386, Apr. 1974.

[75] F. Koch, F. Marschall, J. Meiser, O. Márkus, A. Faisal, T. Schröter, P. Meyer, D. Kunka, A. Last, and J. Mohr, "Increasing the aperture of x-ray mosaic lenses by freeze drying," *J. Micromechanics Microengineering*, vol. 25, no. 7, 2015.

[76] M. Guttmann, B. Matthis, S. Manegold, M. Hartmann, and E. Walch, "Erprobung eines neuen Hartnickel-Elektrolyten für die Mikrogalvanoformung," *Galvanotechnik*, vol. 100, pp. 2120–2126, 2009.

[77] N. Dambrowsky, "PhD thesis: Goldgalvanik in der Mikrosystemtechnik," Forschungszentrum Karlsruhe, 2007.

[78] M. Amberger, "From Li to Ga, Phd thesis, in preparation," Karlsruhe Institute of Technology.

[79] M. Despont, H. Lorenz, N. Fahrni, J. Brugger, P. Renaud, and P. Vettiger, "High-aspect-ratio, ultrathick, negative-tone near-uv photoresist for MEMS applications," *Proc. IEEE Tenth Annu. Int. Work. Micro Electro Mech. Syst. An Investig. Micro Struct. Sensors, Actuators, Mach. Robot.*, 1997.

[80] P. Rai-Choudhury, *Handbook of Microlithography, Micromachining, and Microfabrication: Microlithography*. 1997.

[81] S. Lemke, "PhD thesis: Charakterisierung modifizierter Negativresiste für die Röntgentiefenlithographie - PhD thesis," Technische Universität Berlin, 2011.

[82] S. Nordt, H. Pasch, and W. Radke, "Method development for Epoxy resin analysis," in *Microsystem Technologies*, 2010, vol. 16, pp. 1347-1351.

[83] S. Nordt, "PhD thesis: Analytik von Novolak-Epoxidharzen für die Resistentwicklung in der Mikrosystemtechnik," Technische Universität Darmstadt, 2010.

[84] L. E. Scriven, "Physics and Applications of DIP Coating and Spin Coating," *MRS Proc.*, vol. 121, p. 717, Feb. 1988.

[85] C. Becnel, Y. Desta, and K. Kelly, "Ultra-deep x-ray lithography of densely packed SU-8 features: I. An SU-8 casting procedure to obtain uniform solvent content with accompanying experimental results," *J. Micromechanics Microengineering*, vol. 15, no. 6, pp. 1242–1248, Jun. 2005.

[86] A. del Campo and C. Greiner, "SU-8: a photoresist for high-aspect-ratio and 3D submicron lithography," *J. Micromechanics Microengineering*, vol. 17, no. 6, pp. R81–R95, Jun. 2007.

[87] J. Kenntner, "PhD thesis: Herstellung von Gitterstrukturen mit Aspektverhältnis 100 für die Phasenkontrastbildgebung in einem Talbot-Interferometer," Karlsruhe Institute of Technology, 2013.

[88] A. Schütz, "PhD thesis: Untersuchungen zum Einsatz des Negativresistmaterials SU-8 in der LIGA-Technik," pp. 1–105, 2004.

[89] S. Tagawa, S. Nagahara, T. Iwamoto, M. Wakita, T. Kozawa, Y. Yamamoto, D. Werst, and A. . Trifunac, "Radiation and photochemistry of onium salt acid generators in chemically amplified resists," *Proc. SPIE*, vol. 3999, pp. 204–213, 2000.

[90] L. J. Guerin, M. Bossel, M. Demierre, S. Calmes, and P. Renaud, "Simple and low cost fabrication of embedded micro-channels by using\na new thick-film photoplastic," *Proc. Int. Solid State Sensors Actuators Conf. (Transducers '97)*, vol. 2, pp. 1419–1422, 1997.

[91] H. Henning Winter, "Gel Point," in *Encyclopedia of Polymer Science and Technology*, Hoboken, NJ, USA: John Wiley & Sons, Inc., 2003.

[92] D. Kunka, J. Mohr, V. Nazmov, J. Meiser, P. Meyer, F. Koch, M. Amberger, J. Schulz, M. Walter, T. Duttenhofer, A. Voigt, G. Ahrens, and G. Grützner, "Characterization method for new resist formulations for HAR patterns made by X-ray lithography," *Microsyst. Technol.*, Dec. 2013.

[93] L. Hahn, "INNOLIGA - Abschlussbericht FÖRDERKENNZEICHEN: 16SV3522," 2010.

[94] J. Meiser, "PhD thesis: Entwicklung und Charakterisierung großflächiger Röntgengitter für die Phasenkontrastbildgebung," Karlsruher Institut für Technologie, 2016.

[95] A. Faisal, "MSc Thesis: Limits of X-ray Grating Geometry and Aspect ratio by Freeze Drying of X-ray exposed resist materials," Karlsruhe Institute of Technology, 2014.

[96] N. C. LaBianca and J. D. Gelorme, "High-aspect-ratio resist for thick-film applications," in *Proc. SPIE 2438*, 1995, pp. 846–852.

[97] C. J. Robin, A. Vishnoi, and K. N. Jonnalagadda, "Mechanical Behavior and Anisotropy of Spin-Coated SU-8 Thin Films for MEMS," *J. Microelectromechanical Syst.*, vol. 23, no. 1, pp. 168–180, Feb. 2014.

[98] M. Haj-Taieb, "PhD thesis: Development of High temperature electrodeposited LIGA MEMS materials," Karlsruhe Institute of Technology, 2009.

[99] M. Teutsch, "PhD thesis: Entwicklung von elektrochemisch abgeschiedenem LIGA-Ni-Al für Hochtemperatur-MEMS-Anwendungen," Karlsruhe Institute of Technology, 2013.

[100] M. Zupan, M. J. Hayden, C. J. Boehlert, and K. J. Hemker, "Development of high-temperature microsample testing," *Exp. Mech.*, vol. 41, no. 3, pp. 242–247, Sep. 2001.

[101] C. Eberl, D. S. Gianola, and S. Bundschuh, "Image correlation toolbox - Matlab script." 2010.

[102] L. Dellmann, S. Roth, C. Beuret, G.-A. Racine, H. Lorenz, M. Despont, P. Renaud, P. Vettiger, and N. F. de Rooij, "Fabrication process of high aspect ratio elastic structures for piezoelectric motor applications," in *Proceedings of International Solid State Sensors and Actuators Conference (Transducers '97)*, 1997, vol. 1, pp. 641–644.

[103] J. Zhang, M. B. Chan-Park, and C. M. Li, "Network properties and acid degradability of epoxy-based SU-8 resists containing reactive gamma-butyrolactone," *Sensors Actuators, B Chem.*, vol. 131, no. 2, pp. 609-620, 2008.

[104] R. Engelke, G. Ahrens, N. Arndt-Staufenbiehl, S. Kopetz, K. Wiesauer, B. Löchel, H. Schröder, J. Kastner, A. Neyer, D. Stifter, and G. Grützner, "Investigations on possibilities of inline inspection of high aspect ratio microstructures," *Microsyst. Technol.*, vol. 13, no. 3–4, pp. 319–325, Dec. 2006.

[105] Y. Nawab, S. Shahid, N. Boyard, and F. Jacquemin, "Chemical shrinkage characterization techniques for thermoset resins and associated composites," *J. Mater. Sci.*, vol. 48, no. 16, pp. 5387–5409, Apr. 2013.

[106] A. W. Snow and J. P. Armistead, "A simple dilatometer for thermoset cure shrinkage and thermal expansion measurements," *J. Appl. Polym. Sci.*, vol. 52, no. 3, pp. 401–411, Apr. 1994.

[107] D. U. Shah and P. J. Schubel, "Evaluation of cure shrinkage measurement techniques for thermosetting resins," *Polym. Test.*, vol. 29, no. 6, pp. 629–639, 2010.

[108] M. Baumann, "MSc thesis: Prozessierung großflächiger Substrate im Rahmen des neuartigen Konzeptes der Bewegten Belichtung," Karlsruhe Institute of Technology, 2016.

[109] F. Rueggeberg and K. Tamareselvy, "Resin cure determination by polymerization shrinkage," *Dent. Mater.*, vol. 11, no. 4, pp. 265–268, Jul. 1995.

[110] J. Hammacher, A. Fuelle, J. Flaemig, J. Saupe, B. Loechel, and J. Grimm, "Stress engineering and mechanical properties of SU-8-layers for mechanical applications," *Microsyst. Technol.*, vol. 14, no. 9–11, pp. 1515–1523, Oct. 2008.

[111] H. Lorenz, M. Laudon, and P. Renaud, "Mechanical characterization of a new high-aspect-ratio near UV-photoresist," *Microelectronic Engineering*, vol. 41–42. pp. 371–374, 1998.

[112] M. Holst, "PhD thesis: Reaktionsschwindung von Epoxidharz-Systemen," Deutsches Kunststoff-Institut DKI, 2001.

[113] J. N. Burghartz, *Ultra-thin Chip Technology and Applications*, 1st ed. New York: Springer-Verlag, 2011.

[114] B. L. Henke, E. M. Gullikson, and J. C. Davis, "X-Ray Interactions: Photo-absorption, Scattering, Transmission, and Reflection at E = 50-30,000 eV, Z = 1-92," *At. Data Nucl. Data Tables*, vol. 54, pp. 181-342, 1993.

[115] "Crystalbond 509, available e.g. at SPI Supplies." [Online]. Available: http://www.2spi.com/item/z05110/crystal-bond/.

[116] "Ohio Carbon Blank, Inc.," 2016. [Online]. Available: https://www.ohiocarbonblank.com/.

[117] M. Walter, "PhD thesis: Charakterisierung der Prozessgüte und Verbesserung der Herstellung periodischer Strukturen mit dem LIGA-Verfahren," Karlsruhe Institue of Technology, 2016.

[118] T. Donath, M. Chabior, F. Pfeiffer, O. Bunk, E. Reznikova, J. Mohr, E. Hempel, S. Popescu, M. Hoheisel, M. Schuster, J. Baumann, and C. David, "Inverse geometry for grating-based x-ray phase-contrast imaging," *J. Appl. Phys.*, vol. 106, pp. 54703-54703–7, 2009.

[119] L. Birnbacher, M. Willner, A. Velroyen, M. Marschner, A. Hipp, J. Meiser, F. Koch, T. Schröter, D. Kunka, J. Mohr, F. Pfeiffer, and J. Herzen, "Experimental Realisation of High-sensitivity Laboratory X-ray Grating-based Phase-contrast Computed Tomography," *Sci. Rep.*, vol. 6, 2016.

[120] F. J. Koch, T. J. Schröter, D. Kunka, P. Meyer, J. Meiser, A. Faisal, M. I. Khalil, L. Birnbacher, M. Viermetz, M. Walter, J. Schulz, F. Pfeiffer, and J. Mohr, "Note: Gratings on low absorbing substrates for x-ray phase contrast imaging," *Rev. Sci. Instrum.*, vol. 86, no. 12, 2015.

[121] T. Roth, L. Helfen, J. Hallmann, L. Samoylova, P. Kwaśniewski, B. Lengeler, and A. Madsen, "X-ray laminography and SAXS on beryllium grades and lenses and wavefront propagation through imperfect compound refractive lenses," in *Proc. SPIE 9207, Advances in X-Ray/EUV Optics and Components IX*, 2014, no. December 2015, p. Z20702.

[122] F. Uhlén, J. Rahomäki, D. Nilsson, F. Seiboth, C. Sanz, U. Wagner, C. Rau, C. G. Schroer, and U. Vogt, "Ronchi test for characterization of X-ray nanofocusing optics and beamlines," *J. Synchrotron Radiat.*, vol. 21, no. 5, pp. 1–5, Aug. 2014.

[123] W. Yashiro, S. Harasse, A. Takeuchi, Y. Suzuki, and A. Momose, "Hard-x-ray phase-imaging microscopy using the self-imaging phenomenon of a transmission grating," *Phys. Rev. A*, vol. 82, no. 4, p. 43822, Oct. 2010.

[124] F. Seiboth, "PhD thesis: Refractive Hard X-Ray Nanofocusing At Storage Ring and X-Ray Free -Electron Laser Sources," Universität Hamburg, 2016.

[125] F. J. Koch, C. Detlefs, T. J. Schröter, D. Kunka, A. Last, and J. Mohr, "Quantitative characterization of X-ray lenses from two fabrication techniques with grating interferometry," *Opt. Express*, vol. 24, no. 9, p. 9168, Apr. 2016.

[126] H. H. Wen, E. E. Bennett, R. Kopace, A. F. Stein, and V. Pai, "Single-shot x-ray differential phase-contrast and diffraction imaging using two-dimensional transmission gratings.," *Opt. Lett.*, vol. 35, no. 12, pp. 1932–1934, 2010.

[127] K. S. Morgan, P. Modregger, S. C. Irvine, S. Rutishauser, V. a Guzenko, M. Stampanoni, and C. David, "A sensitive x-ray phase contrast technique for rapid imaging using a single phase grid analyzer.," *Opt. Lett.*, vol. 38, no. 22, pp. 4605–8, 2013.

[128] P. Zhou and J. H. Burge, "Analysis of wavefront propagation using the Talbot effect.," *Appl. Opt.*, vol. 49, no. 28, pp. 5351–9, Oct. 2010.

[129] C. Kottler, C. David, F. Pfeiffer, and O. Bunk, "A two-directional approach for grating based differential phase contrast imaging using hard x-rays.," *Opt. Express*, vol. 15, pp. 1175–1181, 2007.

[130] F. Zernike, "Beugungstheorie des Schneidenverfahrens und seiner verbesserten Form, der Phasenkontrastmethode," *Physica*, vol. 1, no. 7–12, pp. 689–704, May 1934.

[131] J. Ye, Z. Gao, S. Wang, J. Cheng, W. Wang, and W. Sun, "Comparative assessment of orthogonal polynomials for wavefront reconstruction over the square aperture," *J. Opt. Soc. Am. A*, vol. 31, no. 10, pp. 2304-2311, 2014.

[132] A. Last, O. Márkus, S. Georgi, and J. Mohr, "Röntgenoptische Messung des Seitenwandwinkels direktlithografischer refraktiver Röntgenlinsen," in *MikroSystemTechnik Kongress 2015*, 2015.

[133] C. Billotte, F. M. Bernard, and E. Ruiz, "Chemical shrinkage and thermomechanical characterization of an epoxy resin during cure by a novel in situ measurement method," *Eur. Polym. J.*, vol. 49, no. 11, pp. 3548–3560, 2013.

[134] A. Last, "Brechzahldekrement mr-X," *Pers. Corresp. 26.02.2015.*

Schriften des Instituts für Mikrostrukturtechnik
am Karlsruher Institut für Technologie (KIT)

ISSN 1869-5183

Herausgeber: Institut für Mikrostrukturtechnik

Band 1 Georg Obermaier
Research-to-Business Beziehungen: Technologietransfer durch Kommunikation von Werten (Barrieren, Erfolgsfaktoren und Strategien). 2009
ISBN 978-3-86644-448-5

Band 2 Thomas Grund
Entwicklung von Kunststoff-Mikroventilen im Batch-Verfahren. 2010
ISBN 978-3-86644-496-6

Band 3 Sven Schüle
Modular adaptive mikrooptische Systeme in Kombination mit Mikroaktoren. 2010
ISBN 978-3-86644-529-1

Band 4 Markus Simon
Röntgenlinsen mit großer Apertur. 2010
ISBN 978-3-86644-530-7

Band 5 K. Phillip Schierjott
Miniaturisierte Kapillarelektrophorese zur kontinuierlichen Überwachung von Kationen und Anionen in Prozessströmen. 2010
ISBN 978-3-86644-523-9

Band 6 Stephanie Kißling
Chemische und elektrochemische Methoden zur Oberflächenbearbeitung von galvanogeformten Nickel-Mikrostrukturen. 2010
ISBN 978-3-86644-548-2

Band 7 Friederike J. Gruhl
Oberflächenmodifikation von Surface Acoustic Wave (SAW) Biosensoren für biomedizinische Anwendungen. 2010
ISBN 978-3-86644-543-7

Band 8 Laura Zimmermann
Dreidimensional nanostrukturierte und superhydrophobe mikrofluidische Systeme zur Tröpfchengenerierung und -handhabung. 2011
ISBN 978-3-86644-634-2

Band 9 Martina Reinhardt
Funktionalisierte, polymere Mikrostrukturen für die dreidimensionale Zellkultur. 2011
ISBN 978-3-86644-616-8

Band 10 Mauno Schelb
Integrierte Sensoren mit photonischen Kristallen auf Polymerbasis. 2012
ISBN 978-3-86644-813-1

Band 11 Daniel Auernhammer
Integrierte Lagesensorik für ein adaptives mikrooptisches Ablenksystem. 2012
ISBN 978-3-86644-829-2

Band 12 Nils Z. Danckwardt
Pumpfreier Magnetpartikeltransport in einem Mikroreaktionssystem: Konzeption, Simulation und Machbarkeitsnachweis. 2012
ISBN 978-3-86644-846-9

Band 13 Alexander Kolew
Heißprägen von Verbundfolien für mikrofluidische Anwendungen. 2012
ISBN 978-3-86644-888-9

Band 14 Marko Brammer
Modulare Optoelektronische Mikrofluidische Backplane. 2012
ISBN 978-3-86644-920-6

Band 15 Christiane Neumann
Entwicklung einer Plattform zur individuellen Ansteuerung von Mikroventilen und Aktoren auf der Grundlage eines Phasenüberganges zum Einsatz in der Mikrofluidik. 2013
ISBN 978-3-86644-975-6

Band 16 Julian Hartbaum
Magnetisches Nanoaktorsystem. 2013
ISBN 978-3-86644-981-7

Band 17 Johannes Kenntner
Herstellung von Gitterstrukturen mit Aspektverhältnis 100 für die Phasenkontrastbildgebung in einem Talbot-Interferometer. 2013
ISBN 978-3-7315-0016-2

Band 18 Kristina Kreppenhofer
Modular Biomicrofluidics - Mikrofluidikchips im Baukastensystem für Anwendungen aus der Zellbiologie. 2013
ISBN 978-3-7315-0036-0

Band 19 Ansgar Waldbaur
Entwicklung eines maskenlosen Fotolithographiesystems zum Einsatz im Rapid Prototyping in der Mikrofluidik und zur gezielten Oberflächenfunktionalisierung. 2013
ISBN 978-3-7315-0119-0

Band 20 Christof Megnin
Formgedächtnis-Mikroventile für eine fluidische Plattform. 2013
ISBN 978-3-7315-0121-3

Band 21 Srinivasa Reddy Yeduru
Development of Microactuators Based on the Magnetic Shape Memory Effect. 2013
ISBN 978-3-7315-0125-1

Band 22 Michael Röhrig
Fabrication and Analysis of Bio-Inspired Smart Surfaces. 2014
ISBN 978-3-7315-0163-3

Band 23 Taleieh Rajabi
Entwicklung eines mikrofluidischen Zweikammer-Chipsystems mit integrierter Sensorik für die Anwendung in der Tumorforschung. 2014
ISBN 978-3-7315-0220-3

Band 24 Frieder Märkle
Laserbasierte Verfahren zur Herstellung hochdichter Peptidarrays. 2014
ISBN 978-3-7315-0222-7

Band 25 Tobias Meier
Magnetoresistive and Thermoresistive Scanning Probe Microscopy with Applications in Micro- and Nanotechnology. 2014
ISBN 978-3-7315-0253-1

Band 26 Felix Marschall
Entwicklung eines Röntgenmikroskops für Photonenenergien von 15 keV bis 30 keV. 2014
ISBN 978-3-7315-0263-0

Band 27 Leonardo Pires Carneiro
Development of an Electrochemical Biosensor Platform and a Suitable Low-Impedance Surface Modification Strategy. 2014
ISBN 978-3-7315-0272-2

Band 28 Sebastian Mathias Schillo
Prozessentwicklung für die Automatisierung der Herstellung und Anwendung von hochdichten Peptidmicroarrays. 2014
ISBN 978-3-7315-0274-6

Band 29 Nicole E. Steidle
Micro- and Nanostructured Microfluidic Devices for Localized Protein Immobilization and Other Biomedical Applications. 2014
ISBN 978-3-7315-0297-5

Band 30 Jochen Heneka
Prozessentwicklung eines industrietauglichen Verfahrens zur Fertigung von vereinzelten LIGA-Mikrobauteilen. 2015
ISBN 978-3-7315-0326-2

Band 31 Seoung-Eun Kim
Konzeption und prototypische Fertigung einer nicht-invasiven mikrofluidischen Plattform für die Elektrophysiologie (NIMEP) zur Zellenanlyse. 2015
ISBN 978-3-7315-0378-1

Band 32 Elisabeth Wilhelm
Entwicklung eines mikrofluidischen Brailledisplays. 2015
ISBN 978-3-7315-0385-9

Band 33 Viktor Pinneker
Entwicklung miniaturisierter Aktorsysteme basierend auf magnetischen Formgedächtnislegierungen. 2017
ISBN 978-3-7315-0500-6

Band 34 Ali Caglar Özen
Novel MRI Technologies for Structural and Functional Imaging of Tissues with Ultra-short T_2 Values. 2017
ISBN 978-3-7315-0657-7

Band 35 Anne Bäcker
Veränderliche 3D Zellgerüstträger auf Cryogelbasis zur Kultivierung von Prostatakarzinomzellen. 2017
ISBN 978-3-7315-0676-8

Band 36 Frieder Johannes Koch
X-ray optics made by X-ray lithography: Process optimization and quality control. 2017
ISBN 978-3-7315-0679-9